ONCE AND FUTURE GIANTS

ONCE & FUTURE GIANTS

WHAT ICE AGE EXTINCTIONS TELL US ABOUT
THE FATE OF EARTH'S LARGEST ANIMALS

SHARON LEVY

OXFORD
UNIVERSITY PRESS

OXFORD
UNIVERSITY PRESS

Oxford University Press, Inc., publishes works that further
Oxford University's objective of excellence
in research, scholarship, and education.

Oxford New York
Auckland   Cape Town   Dar es Salaam   Hong Kong   Karachi
Kuala Lumpur   Madrid   Melbourne   Mexico City   Nairobi
New Delhi   Shanghai   Taipei   Toronto

With offices in
Argentina   Austria   Brazil   Chile   Czech Republic   France   Greece
Guatemala   Hungary   Italy   Japan   Poland   Portugal   Singapore
South Korea   Switzerland   Thailand   Turkey   Ukraine   Vietnam

Copyright © 2011 by Sharon Levy

Published by Oxford University Press, Inc.
198 Madison Avenue, New York, New York 10016
www.oup.com

Oxford is a registered trademark of Oxford University Press

Library of Congress Cataloging-in-Publication Data

Levy, Sharon, 1959-
  Once and future giants : what Ice Age extinctions tell us
about the fate of Earth's largest animals / Sharon Levy.
    p. cm.
Includes bibliographical references and index.
ISBN 978-0-19-537012-6 (hardcover : alk. paper) 1. Extinction (Biology)
2. Paleontology–Pleistocene. 3. Wildlife conservation. I. Title.
QE721.2.E97L48 2011
576.8'4–dc22   2010025763

ISBN-13: 978-0-19-537012-6

1 3 5 7 9 8 6 4 2
Printed in the United States of America
on acid-free paper

To Hugh, a very major dude

CONTENTS

ILLUSTRATIONS

PROLOGUE

We live surrounded by the ghosts of giants. Mastodons once foraged in Manhattan and trod along the banks of Lake Michigan, where Chicago's skyscrapers now stand. Bear-sized beaver dammed North American streams; herds of camel and horse roamed the plains. In Europe, mammoths shared turf with giant deer, whose massive antlers measured more than ten feet across. Woolly rhinoceros stampeded through Moscow. And *Diprotodon*, a buck-toothed, rhino-sized marsupial, ambled through Sydney and Perth.

Our ancestors knew these great beasts intimately. *Homo sapiens* evolved in a world thronged with monster mammals; Stone Age Europeans painted portraits of woolly mammoth, cave lion, cave bear, giant deer, wild horse, and ox. Ancient artists hunted these creatures, relished their meat, carved tools from their bones—until the giants suddenly disappeared.

Worldwide, about ninety genera of large mammals went extinct during the last 50,000 years. That amounts to more than 70 percent of America's biggest species, and more than 90 percent of Australia's.[1] (A genus is a group of similar species, closely related in evolutionary terms. Domestic dogs, wolves, coyotes, and jackals all belong to the genus *Canis*, for example, along with many now-extinct species.) Paleontologists long assumed that these animals were driven over the brink by a rapidly warming climate, their habitats shrinking as glaciers receded at the close of the last Ice Age. But in recent years a passionate debate has sprung up over the cause of these extinctions as flaws in the traditional scenario have been uncovered. If climate change was the reason for the global die-off, why, for example, did North American camels, ground sloths, and mastodons succumb to the heat this time, having survived so many previous interglacial warm spells? Why did America's mammoths last thousands of years longer than Australia's giant kangaroos? The skeptics pointed to a new culprit: humans.

A growing number of researchers are now uncovering evidence that our ancestors played a pivotal role in the demise of these and other giant animals, known collectively as the megafauna. In North America, mass extinctions of mammoth, mastodon, American camel, horse, lion, saber-toothed cat, and dire wolf coincided not only with the end of the last glacial peak of the Ice Age, but also with the arrival of Clovis people. Members of the first human culture to spread quickly across the continent, Clovis hunters left their elegant stone spearpoints buried in the ribs of mammoths and mastodons scattered from Arizona to Maine. Could overhunting by America's early people have driven the megafauna into oblivion? How could humans armed only with spears have obliterated whole populations of massive beasts?

The answers to such questions might seem irrelevant in a world plagued by a host of more urgent-sounding problems. But lessons from the Pleistocene—the two-million-year epoch of the lost giants—may be critical to the conservation of megafauna living today. Elephants, tigers, gray wolves, grizzly bears, caribou, and kangaroos now face an array of troubles that echo the last days of the mammoth and mastodon. Booming human populations amplify today's threats, which include rapidly shrinking wild habitats, intense human hunting, and global warming at a rate unprecedented in the long span of mammalian evolution. Our ancestors witnessed, and perhaps caused, the decline and fall of prehistoric giants. The fate of today's megafauna is even more closely tied to human actions.

These modern parallels attract researchers to study the ancient extinctions. But there is another, more basic reason: the vanished giants were, to use a technical term, awesome. The thick, rugged bones of Harlan's ground sloth—a vegetarian that once lumbered across the American West—make those of a Tyrannosaurus seem graceful. The broad femur with its bulbous head looks more like a shillelagh than a leg bone. In life this ponderous beast stood more than six feet tall, weighing about 1,500 pounds—its sheer mass protecting it from the attacks of American lions.

Fig. 1   The skeleton of Harlan's ground sloth, a 1,500-pound vegetarian that once lumbered across the American West. Photo courtesy of Page Museum and Los Angeles Natural History Museum.

These great cats were built like their living African relatives, but—as a reconstructed skeleton shows—on a scale 25 percent bigger. The inert bones alone evoke a shiver of primal fear.

I first began reporting on the Pleistocene extinctions in 1999, when a paper published in *Science* claimed that the demise of Australia's giants had been triggered by ancient Aborigines wielding fire sticks.[2] The intensity of the resulting controversy, and its obvious relevance to pressing issues in modern ecology and conservation, captivated me. So did the players in the debate. Visiting with them was like listening in on a group of school-age boys as they obsessed together over the details of a model train, the design of its track and its hand-painted landscape. Even better, the megafauna fanatics were not building a simple diorama of a frozen moment in civilization. They were conjuring an entire world, compiling an understanding of the pulsing, snuffling, utterly wild era of the vanished giants. These guys (the vast majority of researchers in the field are male) overflow with passion for their subject, and their enthusiasm is infectious.

Fig. 2 The saber-toothed cat used its mighty canine teeth to slash and kill horse, camel, and baby mammoth. Saber-tooths likely lived in packs and showed strong family loyalties. Several cats whose remains were found at the famous La Brea tar pits, in what is now Los Angeles, had suffered severe broken bones but survived; this would have been possible only if pack members shared food with them while they recovered. (Photo from Wikimedia commons, en.wikipedia.org/wiki/File:Smilodon_head.jpg)

Working on this book gave me license to visit their labs and field sites, hold mastodon molars and mammoth tusks in my hands, examine dung balls left by giant ground sloths, and meet a leather-jacketed researcher who opens letters with the deadly canine of a saber-toothed cat. (The arc of the ancient tooth is edged with fine serrations, which slice smoothly through twenty-first-century paper.)

All this made terrific fodder for my own inner paleo-child and underscored the power of our fundamental connection to these great creatures. Growing up in the concrete maze of Chicago, I always yearned for some contact with the wild. My favorite exhibit at the Field Museum of Natural History was a mounted mastodon skeleton. I dreamed of slipping past the velvet barrier ropes, shimmying up between the great ribs, and curling myself into a ball in the place where the giant's heart had once beat. I imagined that if I ever summoned the nerve to do this, the pavement outside would crack open and ancient elephants would rise from the earth to tramp along Lake Shore Drive, causing a marvelous panic. Decades later, I found myself talking with expert scientists who shared my childhood fantasy—and who believed it might be made real.

For me, and for many of the researchers working to understand the fate of the vanished megafauna, it's impossible to handle the old bones without mentally clothing them in muscle, sinew, and fur and envisioning the world their owners knew. It was the world in which our own species came of age: modern humans evolved and spread out of Africa over the course of the Pleistocene, penetrating to every corner of the planet. The great creatures we encountered along the way shaped us, whether we feared them as predators or stalked them as prey. Thousands of years later, the giants still radiate animal magnetism, though all that's left of them are skeletons and desiccated dung.

Some researchers, not satisfied to recreate the lost world of the Ice Age only in their minds, support a more drastic step: returning camels, elephants, and cheetahs to their prehistoric American home—in the flesh. In 2005, *Nature* published a manifesto signed by a dozen prestigious scientists, including Paul Martin, an inventive paleoecologist who had been cultivating a mad passion for the Pleistocene since he began his graduate work in the 1950s. Their radical new concept: to move endangered Old World beasts to open spaces in the western United States as ecological stand-ins for their extinct native brethren.[3] The authors included pioneering biologists who, earlier in their careers, had agreed with the conventional wisdom that introduced species are a human-induced cancer on natural ecosystems. Yet here they were, proposing an intentional transfer of exotic African and Asian wildlife to America in the name of conservation. What had changed?

Proponents of Pleistocene rewilding have begun to question traditional ideas about what is, or is not, natural. American conservationists had long assumed that the creatures roaming the landscape in 1492, when Columbus arrived, represented the original, pristine state of the continent. That vision ignores the splendid menagerie of giant mammals that had vanished thousands of years earlier. The Pleistocene giants built up, and sometimes tore down, the forests and grasslands that early people walked. Ecologists have come to understand that humans are not the only creatures that shape their surroundings. Modern bison, tapir, elk, and rhino enrich the soil with their waste and encourage new plant sprouts by swallowing seeds and redepositing them in fertile ground. By bashing and breaking trees, African elephants transform woodland into open savannah. Extinct forms of bison, camel, horse, woolly rhino, and woolly mammoth affected Pleistocene landscapes in similar ways.

The long fossil record of the Pleistocene shows that mammoths and other giants were resilient, surviving many shifts from intense cold to relative warmth. But that was in a world not yet ruled by people, with open landscapes and plenty of room to move: north when the glaciers retreated, back south when they advanced. Today the planet's last wild megafauna live trapped on islands of habitat isolated in an ocean of humanity. In these fragments of wilderness, research shows that the big beasts shape their habitats in many ways, increasing biodiversity and even sheltering fellow creatures from the impacts of global warming. The desire to keep these creatures alive is not a romantic whim. Solid scientific evidence shows that megafauna make the planet a richer, more stable place, with ecosystems more likely to survive rapid environmental change.

Twenty-first-century people need to understand the mammoth and the saber-toothed cat, not to raise musty bones from the dead but to help keep alive what wildness we still have with us. Researchers worldwide are using evidence of defunct megafauna to explore conservation strategies, working at sites from Australia to New York, from the African savannah to the tundra of Greenland. The scientific evidence on how to share the earth with modern megafauna—and what could happen if we fail—is still emerging, driven by researchers whose visions of the distant past shape a very modern debate.

This book tells the story of the megafauna and us. It is a tale of human coexistence and clashes with giant animals, past and present, and our responsibility toward them in the future.

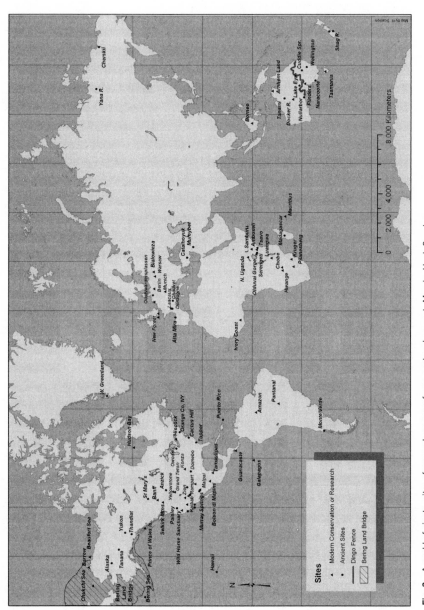

Fig. 3 A world of giants: sites of research on megafauna past and present. Map by Hugh Scanlon.

ELEGY FOR THE MASTODON

READING THE ASHES

Guy Robinson stands neck-deep in a hole he's dug in the dark earth of a woodland in southeastern New York. To reach back to the time of the first humans to walk here—people who knew and hunted 16,000-pound mastodons—he uses a long auger to drill beneath his muddy feet. The tubular handle hugged to his chest, Robinson, a paleoecologist at Fordham University, hauls up a cylinder packed with wet bits of sedge and cattail, like those you might find in any modern wetland. Holding out a fistful of soggy vegetation, he looks as happy as a kid excavating a cookie jar. "This," he says, "could be mastodon fodder." These plants date to 14,000 years bp (before present), when they risked being trod upon or eaten by the last of North America's native elephants.

Robinson chose this forested spot because here, surrounded by the onion fields and housing developments of Orange County, in the Hudson Valley, the earth has remained undisturbed by plows. The walls of his expertly dug hole show a cross-section of the peat layers laid down tens of thousands of years ago, when this ground made up the floor of a broad, shallow marsh filled with glacial meltwater. Alders and oaks grew along the shoreline, and a spectacular array of beasts fed and watered there: native horses and camels, giant deer and beaver, and the best known local giant—the mastodon. (Shorter, stockier cousins of African elephants and mammoths, mastodons once ranged from Alaska to Brazil.) These magnificent creatures, known collectively as the megafauna, vanished 13,000 years ago at the close of the Pleistocene epoch. Robinson's research attempts to discover what caused such widespread animals to die out all at once.

Massive glaciers covered much of northern Asia, Europe, and North America during the long span of the Pleistocene. The epoch, which stretched from 2.5 million to 10,000 years bp, is often referred to as the Ice Age. It was a time of dramatic climate shifts. During the coldest phases, glaciers extended as far south as the 40th parallel, which passes through such now-temperate spots as Oregon and Spain. Periods of intense cold were broken up by warm spells called interglacials, when the climate was similar to today's. We are now living through an interglacial, labeled by paleontologists as the Holocene epoch.[1]

Conventional wisdom long held that the megafauna fell victim to a warming climate at the end of the last glacial peak of the Ice Age. According to this theory, rising temperatures led to changes in vegetation, altering habitat in ways that proved fatal to many large herbivores and in turn to the dire wolves, American lions, and saber-toothed cats that had preyed on them. Today many scientists believe ancient people were responsible for the extinctions, an idea raised with dramatic flair by paleoecologist Paul Martin.

In the 1960s, Martin, then a maverick starting out his career at the University of Arizona, began to talk about what seemed like an outlandish new theory—that Stone Age hunters, who colonized the Americas as the Pleistocene came to its close, had wiped out the continent's mammoths, mastodons, and other big beasts. In Martin's scenario, the naive giants died out quickly, defenseless against the invasion of human predators. The culprits were Clovis people, descendants of East Asians who had honed their hunting skills on Siberia's woolly mammoths and rhinos for generations. Martin described their confrontation with American megafauna as a "blitzkrieg." In his view, human-induced wildlife extinction is nothing new.

Robinson believes he's found evidence to support Martin's overkill theory in the ancient muck of southeastern New York. For him, Orange County is not just another rural patch surviving beyond the sprawl of New York City's suburbs: it is mastodon country, a place where giants once roamed and where a scientist's imagination can still run wild. As we drove past farm fields, his aging 1988 Oldsmobile rumbled over an entire mastodon skeleton, still at rest beneath the pavement of Pumpkin Swamp Road because there were no funds to excavate it. Robinson doesn't need the bones: he uses microscopic bits of preserved pollen, fungal spores,

Fig. 4 Paleoecologist Guy Robinson facing a mounted mastodon skeleton. Robinson uses microscopic fossil pollen and scraps of ancient charcoal to track the fate of New York's mastodons and other extinct giants. (Photo by Masood Kamandy.)

and ash to delve into the Ice Age megafauna's mysterious demise. Putting to work the muscle he developed during his earlier career as a carpenter, he gathers specimens on his own, using only a shovel and an auger.

Robinson's study is the first attempt to track both the arrival of humans and the disappearance of large herbivores in a single region of North America. He argues that the patterns of change in sediments taken from deep beneath the surface of the Orange County's rich glacial peat—known to locals as the Black Dirt—show that populations of mastodons and other large herbivores all collapsed at the same time, around 14,000 years bp. In his reading, New York's first human beings caused this sudden disaster. While some agree that Robinson has found a brilliant new way of confirming that our ancestors were responsible for the death of Ice Age giants throughout the Americas, others warn that he, like many before him, sees in the evidence only what he already believes.

The Black Dirt is ideal for farming and also, it turns out, for preserving fossils. Plows working local fields have been turning up the bones of mastodons, extinct stag moose, and other long-gone monsters for generations. The first mastodon remains found in America, a set of weighty molars, were discovered nearby, and the first near-complete mastodon skeleton ever discovered—a treasure coveted by then-president Thomas Jefferson—was excavated from a farm field outside Newburgh, a town Robinson and I passed on our way to this dig site.

From the beginning, people have seen what they wanted to see in the bones of America's extinct monsters. The devout seventeenth-century colonists who found that first pair of mastodon molars were convinced they had discovered the remains of a human giant, proof that the David and Goliath story was true. Later, Jefferson would treasure big fossils as evidence of his country's boundless potential, in a time when European intellectuals claimed that the native creatures of the New World, including American Indians, were made small, weak, and stupid by a malign climate. One particularly avid proponent of this concept of "American degeneracy," Abbé Raynal, theorized that even Europeans dwindled physically and mentally under the dread influences of wild America. "One must be astonished," Raynal wrote, "that America has not yet produced one good poet, one able mathematician, one man of genius in a single art or a single science."[2]

Jefferson sent mastodon bones across the Atlantic, along with the stuffed body of a modern moose (and Jefferson's own erudite self) to counter this idea. He was predisposed to imagine the fossils he collected as evidence of glamorous beings unique to America. Still, his fascination with the ancient bones survived their failure to verify some of his cherished fantasies. In 1797, Jefferson was brought a find from a West Virginia cave: a massive humerus, coupled with long lower arm bones that ended in a menacing claw 20 centimeters long. Jefferson assumed it was the paw of a gigantic lion, more impressive than any known from the Old World. Later, he came across an illustrated study by the great French comparative anatomist Georges Cuvier, which showed that the claw belonged to a towering but inelegant species of extinct sloth. Undeterred by this error, Jefferson held to the belief that creatures such as mastodons and giant sloths still roamed the unexplored American West and directed Lewis and Clark to keep a sharp eye out when he sent them off on their expedition to the Pacific coast.

Big fossils and ancient human artifacts are scarce. Each discovery holds more significance for its rarity and is subject to Jeffersonian misreadings. Robinson believes that by relying too much on conventional archaeological finds of bones and spearpoints, researchers miss vital evidence. Instead, he has turned to miniscule but widespread markers left behind in ancient soil. The fossil spores of *Sporormiella*, a fungus that grows only on the dung of large herbivores, is common in most sediments from the Pleistocene. By showing that the spores disappear from the sedimentary record right before a burst of microscopic charcoal marks what he contends were man-made fires, Robinson has made an intriguing new argument indicting people for the extinctions.

Robinson spends much of his time working at his laboratory in Fordham's Bronx, New York, campus. The cupboards hold compelling souvenirs, like the baggie that protects a wad of ancient leaves dried to a parchment-like brittleness—the gut contents of a mastodon found more than a century ago in an Orange County field. Bent over his desk, his lean face glowing in the dim light cast by his microscope, Robinson counts the tiniest of fossils. With a glance at a slide, he can tell what kinds of plants grew near the swamps of Orange County thousands of years ago. Alder pollen resembles a squashed pentagon; oak pollen is shaped like the blade of a window fan, its surface bumpy as an orange peel.

The study of preserved pollen makes it possible for modern researchers to explore long-extinct ecosystems. And because the shifting patterns in dominant plant communities over time—from conifer to oak forest, for instance—are well known throughout the northeast, Robinson can use pollen to place his samples in time.

Like most paleontologists studying the late Pleistocene, Robinson uses radiocarbon dating to determine the age of his samples. All living things absorb carbon from the atmosphere, and a percentage of atmospheric carbon consists of the unstable radioactive isotope, carbon-14. This form of carbon spontaneously decays, turning into nitrogen. When a plant or animal dies, it begins to lose carbon-14 to radioactive decay at a steady rate. The amount of carbon-14 present in an ancient bone or leaf can be used to calculate its age, but the technique has some important limitations. Fossils more than 40,000 years old have already lost much of their carbon-14, and it is difficult or impossible to accurately date them using radiocarbon. And because the amount of carbon-14 present in the atmosphere has shifted over the ages, radiocarbon dates often don't match up with calendar years. In the time frame during which North America's last monsters died out and Clovis culture spread across the continent, radiocarbon dates come out about 2,000 years younger than calendar years. Robinson's approximately 12,000 year-old radiocarbon date for the crash in dung fungus populations in New York soils, for instance, translates to about 14,000 calendar years before the present.[3] Some researchers continue to report their data in uncalibrated radiocarbon years, but the dates cited in this book are in calendar years unless otherwise noted.

Robinson has found two dramatic shifts in the microfossil record that may explain the fate of New York's megafauna. First, *Sporormiella* vanishes. Soon after, levels of microscopic charcoal—the legacy of landscape fires—increase more than tenfold. As Robinson reads this evidence, local populations of mastodons and other big herbivores crashed when the first people arrived and found the naive animals easy prey. With most of the monster vegetarians wiped out, fuel built up on the landscape. So when the newcomers lit fires, as hunter-gatherer peoples have always done, they burned hotter and spread farther than ever before.

Both of these shifts took place well before the last major climate changes that marked the end of the Ice Age, supporting Robinson's argument that people, not climate, killed off the megafauna. What's more, when the big herbivores vanished from southeastern New York, the pollen record shows no significant shift in local vegetation type, as would be expected during a dramatic climate change.

At Appleman Lake, Indiana, other researchers have documented the same pattern of *Sporormiella* decline followed by a spike in charcoal; these changes preceded a dramatic shift in vegetation.[4] The megafauna decline there dates to the same window of time that Robinson tracked in his samples of New York muck. At sites scattered from Colorado to California, ancient sediments tell the same story. Large herbivores dwindled abruptly, causing *Sporormiella* to disappear from the sedimentary record. Soon afterward, landscape fires became more widespread. Significant levels of *Sporormiella* don't show up again until historic time, when domestic livestock began grazing the western range, providing new habitat for the fungus.

Researchers in both camps—those who argue that the extinctions were caused by climate as well as those who blame human invaders—are impressed with the use of dung spores to track the disappearance of the megafauna. In the western United States, spore counts plummet at about 13,000 years bp—tantalizingly close to the time Clovis artifacts first appear in the archaeological record. In the east, however, the megafauna crash seems to have happened about 1,000 years earlier. That bit of evidence could change long-held ideas on how the last days of North America's giants played out.

Paul Martin assumed that humans had arrived not long before the last of the megafauna died out about 13,000 years ago. That seems reasonable, given that radiocarbon dating places the youngest megafauna bones and the artifacts of Clovis culture in that window of time. But Robinson's microfossils alter the conventional overkill scenario. "The main populations collapsed well before the last of the animals died out," he says. "The process didn't happen as quickly as in Paul's original blitzkrieg model."

Robinson agrees that most of the big beasts died out quickly, hunted by human settlers. The youngest known fossils, in his view, are the bones of a lonely few creatures that survived for a millennium beyond the heyday of their kind. Long before these last individuals died, their species were doomed. Healthy populations of giant herbivores shape the landscapes that sustain them. But the mastodon at the close of the Pleistocene was so rare that it was rendered environmentally insignificant—a form of biological limbo Robinson calls "functional extinction." Some of today's highly endangered megaherbivores, such as the Javan rhino, suffer the same condition.

Robinson's data on dung fungus spores have reset the clock for mass extinctions in New York. But convincing his colleagues that humans were walking the Hudson Valley 14,000 years ago, and industriously wiping out the megafauna there, is a tougher proposition. There is scant archaeological evidence available to help date the arrival of the first people in the northeast, and what little has been found is cryptic.

The signature tools of the 13,000-year-old Clovis culture have been found at sites in Vermont, Maine, and western New York. Archaeologists working at a few controversial digs in the eastern United States claim to have found human artifacts 18,000 years old. Spearpoints from Dutchess Quarry, just up the valley from Robinson's study sites, are ancient.

But they don't match the style of the classic Clovis points, so placing them in time is problematic; researchers so far have been unable to date them. Robinson argues that microfossils can fill in the gaps. If his interpretation of the shifts in microscopic charcoal concentrations is right, humans had a dramatic effect on the ecology of the Hudson Valley more than a thousand years before the time of the oldest known Clovis artifacts.

Using charcoal as a marker for human arrival might seem like a huge leap of faith: fire was a part of North America's ecology long before human hands held a flint. Indeed, some paleontologists accuse Robinson of reading too much into ancient remnants of charred grass. But he and his mentor, conservation biologist David Burney, have seen the same pattern before in such far-flung spots as Madagascar, Hawaii, and Puerto Rico. The argument that humans left a signature written in ash as they settled new places relies on evidence from islands. And even the most dedicated believers in the climate theory of North America's megafauna extinction cannot deny that prehistoric people wiped out island creatures around the world.

Many of these island blitzkriegs took place in relatively recent times. Madagascar, off the east coast of Africa, was filled with giant lemurs, pygmy hippos, giant tortoises, and flightless elephant birds until about 2,000 years ago, when sailors from Malaysia settled the island. Evidence that dates human arrival includes radiocarbon ages on human-modified leg bones of extinct hippos, the pollen of hemp and other plants introduced to the island by people, and studies of the degree to which the Malagasy dialect differs from its closest linguistic relatives, spoken in the highlands of Borneo.[5]

Madagascar has always been an island of fire. Burney, the director of the National Tropical Botanical Garden on Kauai, has studied Madagascar's prehistory for decades, using preserved pollen and microscopic charcoal to reveal the presettlement landscape: a fire-adapted mosaic of grassland and forest. Within 200 years of human arrival, dung fungus spores, the marker for big herbivores, vanish. Soon after, the amount of microcharcoal deposited in the sediments jumps tenfold and stays high for several centuries. According to Burney, the demise of large herbivores left more flammable brush standing, leading to an intensification of wildfires. Robinson has found the same sequence recorded in the sediments of southeastern New York, though it is 10,000 years older. An abrupt drop in *Sporormiella* counts is followed by a ten-fold spike in charcoal. To him, the pattern is a reliable sign of human arrival.

Based on similar changes in microcharcoal deposits in a lagoon in Puerto Rico, Burney has suggested that people arrived there about 5,000 years bp, two millennia before the time of the first archaeological evidence of human habitation.[6] That ties in neatly with recent evidence that ground sloths, armadillos, and tortoises disappeared from the island around the same time.[7] Burney argues that the accumulation of microscopic charcoal, which lofts high and spreads wide during landscape fires, can indicate human presence more reliably than do cultural artifacts, which are always thinly distributed and hard to find.

Robinson and Burney's microfossil studies strengthen evidence of what some researchers call a "deadly syncopation": the mass extinction of animals that disappear as soon as that ultimate invasive species, *Homo sapiens*, reaches their habitat. As our ancestors spread out of their original home range in Africa and Eurasia, they found a new world of unwary prey. The consequences for such naive creatures have been documented in the relatively recent past and in ancient prehistory. Charles Darwin wrote about watching a young boy, one of the first human visitors to the Galapagos, methodically slaughter hundreds of finches by simply whacking them with a stick as they approached him. In 1741, the German naturalist Georg Wilhelm Steller recorded the way a crew of Russian sailors, stranded on Bering Island off the Alaskan coast, hunted sea cows—by paddling up to an unresisting animal, jabbing a harpoon into it, and dragging it ashore. Steller's sea cow went extinct after less than three decades of human hunting.

Vanuatu in the Southwest Pacific was once populated by massive land turtles sporting head spikes and heavy, club-like tails. People who reached the island about 3,000 years bp exterminated these impressive beasts in only 300 years.[8] A thousand years ago, Polynesian settlers, along with their imported dogs and pigs, wiped out Hawaii's heavy-set, flightless geese. Soon after, the ancestors of modern Maoris reached New Zealand and killed off a population of an estimated 160,000 ostrich-like moas, driving an entire genus of birds to extinction in less than a century.

This rapid destruction was possible because moas, like sea cows and most other large animals, were slow breeders. The human onslaught left them no time to replace themselves—or to evolve a strategy to cope with their new predator.

The phenomenon of island blitzkriegs is well studied and generally accepted by the scientific community. The disagreement comes in as scientists struggle to understand what happened when the first people reached new continents. It's astonishing to think that in New Zealand's rugged (though limited) terrain, a group of 1,000 or so humans could find and exterminate every moa within a few decades—a conclusion supported by numerous finds of butchered moas, all of which date to a small window of time.[9] These sites have been relatively easy to find because they contain such massive, obvious bones—the femurs of the largest moas were as long as the average archaeologist is tall.

Much harder to accept is the idea that in the vast and varied expanses of Australia or the Americas, scattered tribes wielding only Stone Age technology could wipe out not one but dozens of species. Australian researchers are engaged in a vigorous argument over the causes of that continent's megafauna extinction. Down Under, the menagerie of vanished giants included a 9-foot-tall kangaroo, a rhinoceros-sized marsupial, a wombat as big as a hippopotamus, and a marsupial lion. The time of the first human settlement, about 45,000–50,000 years bp, seems to coincide with the disappearance of many of these creatures. In North America, many of the big beasts died out at the same moment that Clovis hunters appeared. But these events also happened near the end of the Ice Age, bringing a major climate change into the mix. The situation has resulted in a long and fierce scientific debate.

To make a career of studying the vanished world of the Pleistocene giants, one has to be in love with them all: the shambling ground sloths that stood taller than an elephant, the towering native American camel, the long-horned bison, the overgrown armadillo that could have masqueraded as a Volkswagen Beetle. So it's no surprise that researchers grow passionate about their findings, or sometimes refuse to consider opposing evidence.

On that hot afternoon in Orange County, I helped Guy Robinson to package up his precious samples of Pleistocene muck. He filled in the deep hole he'd dug, and we loaded his gear in the trunk of the Oldsmobile and started the long drive back toward the city. We had started out early in the morning, and now the sun was setting, but Robinson still made a detour. There was someplace I should see, he said, especially now when the lighting was perfect.

We rounded a curve on a back road and came on a vista: acres of open pasture populated with free-roaming beasts. A camel rolled on its back in the dust, scratching an itch. Two horses nipped at each other and chased off into the distance. An elephant loomed, alone in the background. Here in mastodon country, a landowner had offered up his field as a refuge for unwanted and abandoned circus animals. Robinson had explained this, but I had spent all day with him, talking about extinct giants, holding their food in my hands. It was easy to see, in the slanting shafts of rosy light, not a suburban field in the twenty-first century, but a window into the Pleistocene.

CONJURING THE GIANTS

When Robinson and his colleagues peer back in time to the Pleistocene, they build on Martin's vision of an America thronged with mammoths, native horses, and long-horned bison—and invaded by spear-wielding big-game hunters. Martin spent more than fifty years conjuring the lives of mammoths and ground sloths from microscopic evidence in ancient dung and soil. Over the course of his controversial career, he grew an unparalleled passion for defunct Ice Age creatures.

Martin grew up obsessed with wildlife, roaming the fields around his Pennsylvania home armed with binoculars and a first-edition copy of Roger Tory Peterson's *Field Guide to Birds*. He quit attending church at age thirteen, because a "Sunday spent out of doors was much more illuminating," and arrived as a freshman at Cornell University five years later, already a skilled naturalist. But in 1950, when Martin was twenty-two, a bout of polio permanently damaged the nerves controlling both his legs. He could walk using a cane, but the free-roaming world of field biology had closed to him. The disease struck just after he returned from a long idyll spent collecting bird and reptile specimens in the cloud forest of Tamaulipas, Mexico.

When I asked Martin fifty-eight years later how this blow had felt, he grinned. "I'm eighty," he said. "I can't remember what anything felt like fifty years ago—not even lovemaking. But the seductions of field work are extraordinary. If I'd still been able to do that, I'd never have learned so much about Pleistocene extinctions. So my handicap opened a door for explorations of near time." The phrase "near time" is used by paleontologists to refer to the shifting panorama of life over the past 50,000 years. In Martin's case it had a double meaning, denoting both the latter days of the Ice Age and the time that was most alive in his bumptious imagination.[1]

As Martin recovered from his battle with polio, the ecology of the Ice Age was emerging as a hot topic. New evidence from ancient plant remains was beginning to transform the way scientists understood the epoch. Ed Deevey, an ecologist at Yale, had spent years using fossil pollen to track the ways habitats had shifted in the northeast with the repeated advance and retreat of the glaciers. Then, in 1950, chemist Willard Libby developed the radiocarbon dating technique. For the first time, it was possible to directly date ancient bits of organic matter. Mud that had settled in bogs and lake bottoms tens of thousands of years ago could be used to clock the many biological changes that occurred during the Pleistocene. Deevey seized on the new technique, starting up one of the first radiocarbon dating labs in the United States.

With the heights of the cloud forest beyond his reach, Martin began adventuring back in time instead. He made his first foray into the Pleistocene using data he'd collected on the distribution of birds and reptiles in Tamaulipas to deduce the ways the Ice Age had shaped ecosystems there, far south of the glacier's edge. Trying to understand the influence of the ancient ice was a new kind of challenge, appealing in its complexity. By 1955, Martin was working with Deevey at Yale, learning to identify and count grains of fossil pollen. The atmosphere at the lab was charged with the excitement of a revolutionary shift in scientific wisdom, and Martin soaked it all in.

Eventually, Deevey and his colleague Margaret Davis would use fossil pollen to demolish a long-standing scientific dogma, the idea that plant communities are destined always to develop in the same successional patterns, birch associated with fir, oak intermixed with hickory. Davis's studies showed that when the glaciers retreated at the end of the Ice Age, plants we tend to see as inextricably linked in modern ecosystems recolonized the north at varying rates, sometimes arriving centuries apart. Instead of a preordained set of trees and shrubs always growing together, each species fended for itself as best it could in a time of extreme environmental change. The same kind of pattern would later be shown to exist among animals as well. There were moments during the Pleistocene when the reindeer and the taiga vole, now known only in the far north, wandered through Mississippi and Tennessee. At times they kept company with armadillos and pocket gophers, creatures whose modern ranges still include the southeastern United States but stop thousands of miles distant from Arctic ice.

One afternoon in the winter of 1957, Martin picked up a catalogue of mammals compiled by the prominent paleontologist George Gaylord Simpson. The list noted every known mammal genus, living or extinct, that had walked or swum on the earth. Simpson covered 250 million years of evolution, recording the emergence and disappearance of thousands of creatures—everything from mice to mammoths to giant wombats.

As a diversion from pollen counting on that snowy afternoon, Martin began to plot mammalian extinctions from the late Pleistocene against those that had taken place farther back in time. He found a remarkable pattern. Over the millions of years of mammal evolution, many species of all sizes and habits had evolved, gone extinct, and been replaced by something else. At the end of the Pleistocene, however, large terrestrial mammals were hit hard, while marine mammals and small land mammals came through almost untouched. In near time, North America lost more genera than it had in the preceding 1.8 million years. Nearly all of the doomed were large mammals: horses and camels, which first evolved in the Americas; mammoths and mastodons; giant ground sloths; glyptodonts, outsized relatives of the armadillo; lost species of musk ox and bison; and big predators like dire wolf, saber-toothed cat, American lion, and short-faced bear. All had survived repeated, dramatic climate shifts, as the glaciers waxed and waned during the long course of the Pleistocene. What deadly force, Martin wondered, had destroyed them when the ice pulled back for the last time?

The pattern of almost exclusively large animals going extinct was consistent on the continents then being settled by humans for the first time: Australia and the Americas. Whales, sea cows, and other marine mammals survived unscathed until historic times. It occurred to Martin that there could be a neat explanation for the near-time die-offs: people did it. Man the hunter had radically altered the face of the planet, long before the invention of agriculture or industry.

Martin was quick to introduce his idea of ancient human overkill at seminars on the Pleistocene. Most of his colleagues laughed the whole thing off: to them, the idea of Stone Age hunters accomplishing such a feat—the destruction of multiple large mammal species—was ridiculous. Martin, unfazed, moved to the University of Arizona in Tucson to continue searching for evidence of his theory. Arizona was an ideal home for him: it had a good radiocarbon lab and was located in arid country likely to contain well-preserved remnants of the Pleistocene.

On his first day at work, Martin's new boss handed him a softball-sized mass of dried plants and dust. At Rampart Cave in the Grand Canyon, researchers had found bones of the extinct Shasta ground sloth, along with layers of its dung so well preserved that the cavern still smelled of manure. Carbon dating proved the feces—including the dung ball Martin held in his hand—to be at least 10,000 years old.

Today, the only surviving sloths are the tree sloths of Central and South America, small and elusive compared to their vanished cousins. In Pleistocene times, ground sloths—large, obvious creatures—ranged from Alaska to the far tip of South America. There were fourteen different genera. The largest, *Megalonyx*, the source of Jefferson's "great claw," was built like a tank, with a massive skull designed for crushing branches, and may have stood taller than a mammoth. The Shasta ground sloth, the creature that had left its traces behind in Rampart Cave, was smaller and lighter but still strikingly weird by modern standards. It had the strange, peglike teeth typical of its family. The toes of its hind paws—its main means of locomotion—curved under the sole of the flat foot, making it difficult to imagine how the animal managed to move at any speed. "One more defect and they could not have existed," commented George-Louis Leclerc, an influential scientist of Thomas Jefferson's day.[2] Yet by Martin's time the fossil record made it clear that ground sloths had flourished alongside mammoths, horses, camels, and other American beasts for millions of years.

Under the microscope, a sample of the dung ball taken from the surface of Rampart Cave's deep deposit of sloth manure proved to be full of spindly white stars—tiny, distinctive hairs that grow on the leaf of the globe mallow, a shrub that remains common around Tucson. Martin wandered into the desert, plucked a globe mallow leaf, and chewed it. "It's not bitter," he remembered, "but it's not something I'd want to eat as a steady diet item. But then, I'm not a ground sloth."

In addition to the mallow hairs, Martin found the pollen of a number of other plants he knew well. At high magnification, the grooved oblongs of *Ephedra* pollen sprang into focus like a row of Goodyear blimps; the angular pollen of the cholla cactus were patterned with regular dents that resembled the face of an alien in a 1950s sci-fi movie. Martin knew he could easily find the same plants the sloths had eaten still growing near the caves where the last of the animals had sheltered. That didn't fit with the climate change theory, which held that sloths and other big beasts had died out from a lack of suitable habitat as the glaciers retreated and temperatures rose.

Rampart Cave held a record of sloth diets reaching more than 25,000 years into the past. This evidence showed that the odd creatures had survived dramatic climate shifts: sagebrush, for instance, which in the 1960s grew only above 4,000 feet, had been common around the cave, at 1,750 feet, during the cold, wet days of the glacial maximum. Depending on the time and the circumstances, sloths had been able to thrive on the globe mallow and cactus that survive in our own hot, arid times, as well as cold-adapted plants like sagebrush and juniper.

The sloth possessed a rare talent: the ability to live on shrubs like creosote and saltbush, desert plants that modern creatures, both native and introduced, leave untouched. The Shasta ground sloth routinely ate snakeweed (*Gutierrezia*) and rabbit brush (*Chrysothamnus*), which are poisonous to sheep and cattle. No other browser came along to fill the empty niche of the ground sloth after its demise. Its extinction, Martin concluded, was an ecological loss without replacement, a rare event in the history of evolution. As he reconstructed the sloth's vanished world, he became more firmly convinced that some long-standing assumptions in paleontology were wrong.

At the time, it was common for researchers to assume that the vanished megafauna had lived only in cold, wet climates and that they had died out due to drought. The presence of megafauna bones was thought to be a reliable indicator of a wet climate at that time and place. That assumption, Martin argued, was false: as far back as Darwin, observant naturalists had pointed out that large mammals can and do live in arid habitats. Now the dung at Rampart Cave provided powerful evidence that ground sloths had been adaptable, able to cope with a wide range of conditions—including the aridity that existed 13,000 years ago, as the Ice Age ended and the first humans arrived in Arizona. "The late Pleistocene archaeological record shows that man and sloth were contemporaneous,"

Fig. 5   Paul Martin with ground sloth dung ball. Evidence Martin found in ancient sloth manure catalyzed his development of the overkill theory of Pleistocene extinctions. (Photo courtesy of Mary Kay O'Rourke.)

he wrote in a paper on the Rampart Cave evidence in 1961, "a circumstance inconclusive in itself. Nevertheless, extinction might become understandable if early man were identified as the main cause. We find no other."[3]

By 1965, Martin was looking to expand his theory of Pleistocene human overkill beyond the Americas. He traveled to Tanzania to visit Louis Leakey, the famous paleoanthropologist who, along with his wife, Mary, had made spectacular discoveries of ancient hominid skeletal remains and tools at Olduvai Gorge. Martin wanted to know more about some of the extinct mammals whose bones had turned up at the gorge—pigs, baboons, and saber-toothed cats.

The stone tools the Leakeys had collected made a profound impression on Martin. The most primitive tool kit consisted of river cobbles broken at one end by heavy blows from a second rock, producing a sharp edge that could be used for cutting and butchering. These Oldowan tools, more than 2 million years old, were made by *Homo habilis*, an ancestor of modern humans whose bones the Leakeys had found at Olduvai.

Millennia later, *Homo erectus*, a creature very similar to modern humans, showed up, bringing with it a new kind of stone tool—one that took real skill to create. Known as hand axes or bifaces, they were worked by percussion flaking to yield a sharp edge along both sides of the tool. Some anthropologists think the bifaces were used for everything from chopping wood and cracking nuts to butchering animals: a Stone Age Swiss Army knife.

What caught Martin's attention was the timing of the new tool kit relative to Africa's own megafauna extinction event. While Africa's stock of large mammals remains gloriously diverse compared to anyplace else in the modern world, Martin calculated that 40 percent of its Pleistocene genera are gone. Most of the large African mammals that died out during the Pleistocene lived long enough to coexist with early humans wielding hand axes. They vanished tens of thousands of years earlier than America's megafauna, during the days of the hand axe culture. As he handled the artifacts from Olduvai, the pieces fell into place in Martin's mind: worldwide, humans had been the driving force behind Pleistocene extinctions. The disaster hit first in Africa, where people had evolved, but most of the megafauna had survived because they had grown up together with this new-fangled predator and had time to adapt. But tens of thousands of years later, when the first people entered North America, they had become expert at making stone tools and carried spears tipped with razor-sharp pieces of chert. These adept hunters walked into a continent full of unsuspecting beasts.

From Tanzania, Martin flew to Madagascar, which would become the smoking gun in his overkill scenario. Only 255 miles of open ocean separates the island from the African continent. If drought indeed drove Africa's late Pleistocene extinctions, as Leakey and most other scientists believed, then it should have hit Madagascar, too. But all the island megafauna lived on, while *Homo erectus* gave way to modern humans and people spread to every corner of the globe. Madagascar was one of the last, most remote spots our species reached. We didn't get there until 2,000 years ago. Soon after, the giant lemurs, pygmy hippos, and elephant birds disappeared.

In October 1966, Martin published his theory in the prestigious journal *Nature.* "The chronology of the extinction—first in Africa, second in

America, finally in Madagascar—and the intensity of the extinction—moderate in Africa, heavier in America, and extremely heavy in Madagascar where it affected much smaller species than on the continents," he wrote, "seems clearly related to the spread of human beings, to their cultural development, and to the vulnerability of the faunas they encountered."[4] Leakey responded with an outraged dissent, quibbling with details of Martin's analysis and dismissing his vision of African overkill. *Nature* published several rounds of debate as the two researchers fired scholarly shots at each other. Martin, the young paleoecologist with the nutty theory, had arrived.

The following year at a congress of the International Association for Quaternary Research, Martin presented his magnum opus, an analysis of the timing of Pleistocene extinctions around the globe. Sorting through a bewildering mass of data, he weeded out long-standing assumptions that no longer made sense. Flawed dating of one fossil find, for instance, had placed the mastodon, a woodland creature, alive on an open plain 8,000 years ago. But a solid combination of evidence from radiocarbon dating of fossil bones and pollen made it clear that North America's last mastodons died out around 13,000 years bp, at the very end of the Ice Age. Some authors had claimed that the last Shasta ground sloth disappeared tens of thousands of years earlier, an idea debunked by Martin's work at Rampart Cave. The technique of carbon dating was still young and prone to many errors in interpretation. Pointing out these mistakes—including some he himself had committed in the past—Martin produced a convincing argument that all of North America's lost giants had died out together, 13,000 years ago, during the brief glory of the Clovis hunting culture.[5] The megafauna vanished over a mere few hundred years, in the geologic blink of an eye. Once the giants were gone, the Clovis spearpoint also disappeared.

Martin's analysis dovetailed perfectly with work by Vance Haynes, a geological archaeologist and a colleague at University of Arizona. Haynes had by then spent years applying the radiocarbon technique to early Paleoindian archaeological sites. It was his work that proved the spectacularly quick rise and fall of Clovis culture and located it firmly in time, at about 13,000 years bp. After that, the ancestors of modern Native Americans lived on, but they began to make different weapons—smaller spearpoints that, both Haynes and Martin suggested, were more suited to hunting bison and smaller beasts that had survived the mass die-off.

"On a world scale, the pattern of Pleistocene extinction makes no sense in terms of climate or environmental change," Martin wrote.[6] "When the chronology of extinction is critically set against the chronology of human migrations and cultural development, man's arrival emerges as the only reasonable answer." And, he noted, he was not the first to suggest that people had finished off the Pleistocene megafauna: no less a luminary than Alfred Russell Wallace, the man who, along with Charles Darwin, had conceived the idea of evolution by means of natural selection, had arrived at the same answer to the mystery. In the 1860s, this had been only a rough guess; back then there had been no way of showing that early human colonization coincided with extinctions. For Martin, radiocarbon dating made all the difference.

The 1967 meeting set off a flurry of new Pleistocene research and polarized the field between advocates of climate-caused extinction and believers in overkill. Martin's vision would shape a number of scientific careers, including those of his long-time opponents in the debate.

Russell Graham, a geoscientist at Pennsylvania State University and a staunch supporter of climate change as the cause of the mass extinctions,

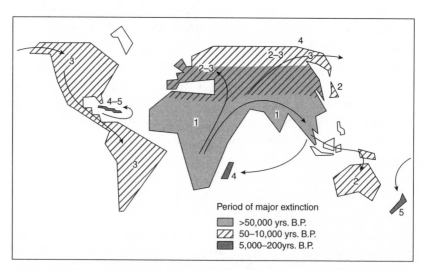

Fig. 6 Map showing Pleistocene era expansion of human populations and timing of megafauna extinctions. Based on Paul S. Martin (2005), *Twilight of the Mammoths*, Berkeley: University of California Press, figure 1.

believes that too many people, scientists and environmentalists alike, uncritically accept the idea that humans bear responsibility for the die-off. Given our species' accelerating destructive impact on animals and habitats worldwide, he says that it's easy to get caught up in a spiral of collective guilt over what our distant ancestors may have done.

Part of the problem with climate-based arguments is that they've become increasingly complex over the last few years. There is no evidence of a dramatic difference between the warming at the close of the Pleistocene and those that took place during earlier interglacials, so a climate-based cause cannot be explained in simple terms. The subtleties of shifts in the fossil pollen record don't provide the same kind of dramatic, easily envisioned story conjured by Martin's overkill theory.

Anthropologist Donald Grayson of the University of Washington—whom Martin cheerfully described as his "arch-rival"—was a graduate student in 1967 when the overkill argument grabbed his attention.

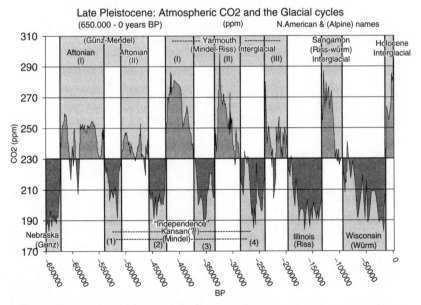

Fig. 7 Glaciers waxed and waned many times during the more than two million year span of Pleistocene time. During peaks of glaciation, ice sheets stretched as far south as Oregon and central Europe. At these times atmospheric carbon dioxide levels dropped. This graph shows changes in atmospheric carbon dioxide, recorded in Antarctic ice deposits over the last 650,000 years. Ancient sea surface temperatures rose and fell in tandem with atmospheric carbon dioxide levels. Data compiled from Dome C, Vostok and Law Dome ice cores and graphed by Tom Ruen, available on Wikimedia commons.

Although he has come to see the model as deeply flawed, he considers it a masterful synthesis, one that only Martin could have created. The idea was not only an elegant interweaving of scientific insights. It was a compelling story, hinging on the simple, dramatic image of a few men taking on a cornered mammoth with only stone-tipped spears—and succeeding.

Grayson began to study the extinctions of North American birds in the late Pleistocene and found a pattern that clashed with the overkill theory, which predicts that only big game animals, or their close ecological partners, should have been hard-hit. He found that the extinction ratio of birds to mammals was very similar. Some of the vanished birds were large scavengers, like the teratorn, an outsized vulture with a 12-foot wingspan. It was easy to imagine that the teratorn, as Martin suggested, was adapted to feed on the carcasses of dead mammoths, camels, and sloths and died out when its food supply vanished. But how, Grayson asked, would human impacts account for the loss of Ice Age species of blackbirds and pigeons, which also disappeared?[7]

Many archaeologists remain deeply skeptical of the overkill theory. Discoveries of sites where mammoths or mastodons were butchered by Clovis people are dramatic but few and far between: today most researchers accept a total of fourteen North American sites as verified elephant kills. There is also some evidence that Clovis-era hunters preyed on native horses and camels. But a few kill sites scattered over the vast expanse of the continent seem to make a weak case for indicting Stone Age people in the mass deaths of hundreds of thousands of large animals. In 1975, when Martin teamed up with computer modeler Jim Mosimann to simulate continental overkill on a clunky IBM mainframe, they produced a scenario of such rapid, widespread slaughter that the lack of butchered fossils, Martin argued, confirmed his theory. It had all happened so quickly, he said, that little evidence had been preserved.[8] "It is a rare hypothesis that predicts a lack of supporting evidence," Grayson would later point out, "but we have one here, and we have it only because evidence for it is, in fact, lacking."

Burney, one of Martin's closest intellectual allies, ultimately proved that some of Martin's ideas about megafauna extinction were wrong. Even on the island of Madagascar, where animals would have had nowhere to run from invading human predators, the process of extinction took much longer than Martin's blitzkrieg model predicted.

The decline of the megafauna led to overgrowth of plants and increased the frequency and intensity of wildfires, which would have affected large animals in unpredictable ways. The last of the island's native hippos and giant lemurs died out 1,500 years after human settlement—not a blitzkrieg, as Martin had envisioned, but a kind of slow fade. Robinson's data from southeastern New York show a parallel set of changes. The particulars may vary from one location to another, but throughout Burney sees one critical common denominator: extinction begins when humans arrive.

The argument is far from over. In 2003, Grayson and his colleague David Meltzer published a scathing criticism of Martin's ideas in the *Journal of Archaeological Science*, titled, "A Requiem for North American Overkill." "Martin's position has become a faith-based policy statement rather than a scientific statement about the past, an overkill credo rather than an overkill hypothesis," they wrote.[9] Yet Grayson, like Burney, holds a deep affection for Martin as a man and a profound respect for his nimble mind.

In "Requiem," Grayson and Meltzer pointed out that Martin's first detailed development of the overkill hypothesis appeared in 1967, the same year the Environmental Defense Fund was launched, five years after Rachel Carson's *Silent Spring* had hit the bookshelves, three years before the creation of Earth Day. The overkill argument, he argued, had captured the popular imagination during a time of intense and growing concern over humanity's impact on nature.

Martin had no problem with the suggestion that his ideas have been shaped by his times. He came of age in an era when naturalists routinely killed the creatures they studied and hatched his dramatic theory just as science began to tally the biological destruction wrought by modern society. When he looked back on his own youth, the seasons spent blasting away at birds in the name of scientific collection, he saw echoes of overkill. If he was the sort of man who indulged in regret, he would have mourned all the hawks and thrushes he'd killed and stuffed.

Martin's ideas have forever changed the way scientists think about megafauna and the Ice Age extinctions, though he himself appeared not to realize the extent to which this is true. The rise of overkill theory has also opened new avenues of research in fields far beyond paleontology. Martin's insights influence researchers studying the ecology of living megafauna, from Kenya to Wyoming, whether or not they know his name. A growing body of research, inspired by Martin's analysis, now suggests that human harvest of large fish and whales in historic times has radically changed the ecology of the sea, creating a marine echo of the late Pleistocene loss of land animals.

Decades later, the Pleistocene overkill theory continues to get plenty of press, and the scientific debate over the causes of the extinctions rages on. Without a time machine, the true nature of the Pleistocene extinctions may never be fully known. Yet clues hidden in mummified balls of dung, in fossil pollen, in scraps of ancient DNA, and in the bones of the long-dead creatures are more than enough to keep researchers hooked. Many scientists remain fixated on the image of the wild giants, unable to stop imagining their gusts of hot breath, the rumble and thunder of mammoths and camels on the move. The earth must have shaken when the big beasts walked. New research now suggests that it changed forever when they vanished. Megafauna shape the world around them, in ways science is only beginning to understand.

LEAVES OF GRASS

In places that suffered a sudden, dramatic loss of megafauna—Australia, New Zealand, Madagascar, the Americas—plants behave in puzzling ways. They produce nutritious fruits, which drop to earth and rot beneath the parent tree, the seeds within destroyed by beetles and rodents. These plants display "megafaunal dispersal syndrome," a phenomenon first described by Paul Martin and tropical ecologist Dan Janzen in 1982.[1] Martin and Janzen, along with a growing number of their colleagues, consider these plants evolutionary anachronisms, relics of ecosystems shaped by the Pleistocene giants.

The only good reason to grow an edible fruit, they argue, is to attract an animal to eat it, pass the seeds, and disperse them in the process. Every plant that reproduces using seeds needs a way to spread them around. Some species produce seeds that float on wind or water, but trees that produce big, heavy fruits rely on big, heavy animals to move the next generation onto fertile ground. Seeds that drop to earth beneath the parent's canopy are doomed: easy targets for predators such as rodents and insects that swarm around fruiting trees. The few that survive to sprout will be shaded to death by the tree that produced them.

During his field work in Costa Rica, Janzen had noticed that many trees in the forests there produced a rain of sweet fruits that dropped to the ground and rotted, largely untouched. On a visit to Uganda, he had the chance to compare the fate of fruits in two sections of forest, one where elephants had been killed off and one where they had been left unmolested. The hefty fruits of the krobodua tree, which enclose large, tough nuts, were devoured where the elephants lived, and young seedlings could be seen sprouting in piles of dung along forest trails. In the area devoid of elephants, the fruits rotted and few new seedlings grew up; the scenario mirrored what he had seen happening for years in Central America.[2] Janzen hypothesized that what seemed a useless adaptation in Costa Rican trees was the artifact of a broken partnership with vanished megafauna: horse, camel, giant ground sloth, and gomphothere, a cousin of the mastodon that roamed Central and South America throughout the Pleistocene.

He tested this idea by feeding pods of the guanacaste tree to horses in Costa Rica and painstakingly tracking the fate of seeds they defecated. The guanacaste, a tree that forms majestic, dome-shaped canopies, packages its seeds in thick pods curved into the shape of a human ear. In places where there are no cattle or horses to eat them, the seed pods lie in useless heaps beneath the parent tree. Janzen found that horses relished the pods and made excellent dispersers for the seeds. Domestic horses were the ideal test animal for this experiment in his view ("escapees from the Pleistocene extinction" that had been conveniently returned to their American homelands by Spanish conquistadors.) The journey through a horse's digestive tract can take as little as two days or as long as two months, ensuring that seeds are widely scattered across the landscape. Dropped in a pile of nutrient-rich dung, far from the hazards of its natal tree, a sprout has every chance to flourish. A fecal coating also offers protection from bruchid beetles, which lay their eggs on seed pods. When the larvae hatch, they burrow inside and devour the seeds.

The megafaunal dispersal syndrome applies to North American plants as well. Near Martin's home in urban Tucson, mesquite pods collect on the sidewalks, their pods pocked with the holes made by bruchid beetles. On a hectic street outside Robinson's laboratory in the Bronx grows a honey locust tree, its trunk and branches studded with forbidding six-inch thorns. The long, sweet fruits are designed to lure the giant herbivores that, in times long past, dispersed the locust seeds; the thorns deterred mastodons and ground sloths from feeding on the rest of the tree.

The honey locust, a tree native to North America, now grows along city streets throughout the United States. Able to thrive in all kinds of soil, tolerant of compacted earth, it is familiar to residents of Chicago and New York. Janzen thinks the species coevolved with mammoth,

Fig. 8 This honey locust tree, growing outside Guy Robinson's lab in the Bronx, is studded with six-inch thorns that evolved to defend the plant against hungry mastodons. (Photo by Masood Kamandy.)

mastodon, and ground sloth and needed its megafaunal partners to spread and flourish. Each seed is encased in a tough coat that must be breached before it can sprout—a way of preserving the seeds until they have passed through the gut of some big animal. Modern horticulturists have described ways of getting a honey locust seed to germinate: soaking it in concentrated sulfuric acid works, as does taking a metal file to the seed coat, or passing it through the digestive system of a cow. As a result of these substitutes, the honey locust is more widespread today than it was during the Pleistocene. Horses and cows brought to the Americas by European settlers, Janzen suggests, replaced absent mastodons and sloths in the life cycle of the honey locust. Devouring its nutritious pods, domestic megafauna helped to expand the tree's range beyond its prehistoric limits.[3]

In a recent study, Brazilian ecologist Mauro Galetti and his colleagues refined Janzen and Martin's original seed dispersal hypothesis, offering the first clear definitions of megafauna fruits.[4] Their classifications are based on the size, shape, color, and texture of large fruits known to be eaten and dispersed by living elephants. Across Brazil's varied habitats—wetland, savannah, rainforest—the researchers found 103 plants that produce fruits that seem adapted for megafauna. These fruits pack a more concentrated load of seeds than those eaten primarily by birds or other small animals. The individual seeds are also larger. This, explains Galetti, reveals an important benefit of luring big animals as dispersers. Large seeds can survive for longer and in tougher environmental conditions than can small ones, because they hold more food reserves for the infant plant. Such seeds can grow even after being partially eaten by a rodent or an insect.

In the South American Pantanal, a broad sweep of wetland and forest that still hosts native tapir and peccary along with domesticated cattle, horses, and pigs, trees produce fruits similar to those found in African reserves where elephant and giraffe still roam. The Brazilian researchers concluded that large frugivores (animals that feed on fruit) drove the evolution of fruit traits on both continents and that numerous domesticated and introduced animals in the Pantanal serve as substitutes for lost Pleistocene megafauna. (Feral pigs proved to be the most prodigious seed dispersers in the Pantanal, more effective than peccary or tapir.)

Megafauna-adapted plants in the Americas have survived the loss of their ecological partners in a variety of ways. Some are dispersed by large rodents, like the rabbit-sized agouti of Central and South America, which collects large seeds and caches them underground. There, protected from beetle attack, many seeds sprout and grow. Others, such as avocado, cacao, and persimmon, have been cultivated by humans. Still other trees have contracted in range and are now found only near water, where their heavy fruits are spread by floating downstream or, as Galetti found for a species of palm in the Pantanal, by the fruit-eating pacu fish. Only the largest fish pass seeds through their gut intact, however, and in a sad echo of the Pleistocene overkill scenario, Galetti points out that commercial fishing for pacu, which focuses on taking the biggest of the fish, could seriously affect the regeneration of palms.[5]

The tight relationship between megafauna and their plant partners is not just a biological oddity. It's a clue that may prove vital to the conservation of modern ecosystems. Native herbivores continue to decline throughout the world's tropics, home to many megafauna-adapted trees. The tapir, a reclusive, fruit-eating forest giant with a rubbery snout, can weigh as much as 1,000 pounds. This relic of the American Pleistocene is threatened by hunting and habitat loss as tropical forests are cut and burned. Ecologist Jose Fragoso has sorted through hundreds of tapir droppings and mapped the distribution of native palm seedlings in the Brazilian Amazon. He found that palm seeds encased in dung were protected from the attacks of bruchid beetles and used aerial photography and satellite imagery to confirm that the patchy pattern of new palm growth matches the location of tapir "latrines" on the ground.[6] Work by Fragoso and others shows that forests suffer from the loss of the tapir and other large frugivores. In a Panama forest, the intensity of poaching correlated with the fate of seeds: where large mammals like tapir and peccary were hunted out, most seeds were destroyed by beetles or rodents and few new trees sprouted.[7] A similar pattern occurred when large native mammals were hunted out of forests in Cameroon.[8]

Seed dispersal is only one of the powerful ways in which megafauna shape their surroundings. Bison on the Great Plains increase the diversity of prairie grasses through selective grazing. By biting off the tender buds of willow and aspen, moose allow spruce trees to grow tall and affect the makeup of North American forests. White rhinos in South Africa alter the grasslands where they graze, controlling the extent of wildfires.[9] Zebra, wildebeest, and gazelle follow each other through the savanna of the Serengeti, sequentially devouring different plant parts and fertilizing the whole ecosystem with their dung and urine, rich in nitrogen that fuels abundant plant growth.[10]

Yet America was once more crowded with fabulous beasts than the modern Serengeti. When humans first stumbled onto the continent, they found a country shaped by mammoth and mastodon. Ancient bones and tusks of American elephant, along with the tools left behind by the people who butchered them, offer a tantalizing glimpse of that vanished world. Combined with studies of living elephants and rhinos, these artifacts provide insights not only into long-ago extinctions, but also into the urgent struggle to save living giants.

MAMMOTH TRACKS

ICE AGE DIARIES

The main work bench in Daniel Fisher's lab at the University of Michigan–Ann Arbor holds a jumble of mastodon and mammoth tusks, some still whole, and others cut open in precise halves like split bananas. Those once carried by adult males form great arcs, 6 feet or more from base to tip. Others, left behind by females or youngsters, are narrow and relatively short, like a collection of scimitars on a graduated scale.

Fisher's beard has gone white, and he examines fossils through round wire-rim glasses. Slender and soft-spoken, he has an eye for the minutiae that other researchers have tended to ignore and a talent for interpreting fine points in novel ways. He has taught himself to read ancient tusks, revealing them to be diaries that depict Pleistocene life in vivid detail. Amid the clutter, he points out the tusk of the Hyde Park mastodon, a massive male from New York's Hudson Valley, who fought many times for the right to mate and died in his mid-thirties of wounds inflicted by a rival. Nearby are bits of the North Java mastodon, a female who lived a full life, bearing and raising six calves before she died at forty. Fisher believes these ancient life histories can help reveal the fate of the last mastodons and mammoths.

Extracting such biographies from the tusks of extinct American and Siberian elephants is a demanding specialty that Fisher fell into by accident, more than twenty years ago. In the early 1980s, not long after he joined the paleontology department at the university, Fisher was called out to a Michigan farm where workers digging a pond had stumbled across a mastodon skeleton. They had stopped work when a bulldozer slammed into the tusks, which broke with a loud cracking sound.

When the young scientist reached the site, he climbed up over a hill of disturbed earth and found fragments of shattered tusk scattered around. Stripes were visible across the broken surfaces, regular light and dark couplets in alternating shades of brown. Fisher realized those patterns might represent years of tusk growth. With growing excitement, he imagined the trove of information he might extract from such growth lines.

Tusks are highly modified incisor teeth. In the same way that human children lose their baby teeth, young mammoths and mastodons had milk tusks that dropped away in infancy to be replaced by a permanent tusk that grew throughout the animal's life. Gleaming within a disk of ivory sliced off of a male mastodon's tusk are concentric circles of light and dark, like the growth rings inside a tree trunk. Tusks cut in half lengthwise reveal a profile of stacked dark and light cones. These lines echo the V-shape of the pulp cavity, which contained the soft tissue that secreted dentin, the stuff of solid ivory. Dark lines formed during the hungry days of winter, when tusks grew slowly and dentin layers were laid down close together. In spring and summer, fresh sprouts of grass and shrubs made foraging easy, and tusks grew faster. The warm season left more widely spaced, and therefore lighter colored, growth lines. In full-grown animals, the tusks of males were much longer and stouter than those of females. It took more pulp to support these living ivory spears, so a male's pulp cavity would leave a much bigger hollow in the tusk than a female's. The tusk of the Hyde Park male, for instance, has a pulp cavity two feet long.

Under the microscope, Fisher discovered daily, weekly, and monthly patterns of growth whose variations mirrored the changing of seasons. Using the lines laid down closest to the base of the tusk, he could determine the season when an animal died. Counting up the annual lines along the entire length of a tusk, he could estimate the length of a life span. But often the most fascinating details lay in between.

Early on in his exploration of tusk records, Fisher was examining several male specimens. At the part of the tusk laid down when these mastodons were about ten years old, he noticed a year's growth increment that was dramatically thinner than the ones preceding it. The years following gradually became thicker, suggesting a famine followed by a slow recovery. In tusks thousands of years old, Fisher had found a record of mastodon social structure that mirrored the descriptions of modern elephants then being written by Cynthia Moss and other biologists working in Africa. Moss had described the fate of every pubescent male: exile from his natal herd.[1] Adult females live in extended family groups, and daughters may stay with their mothers and aunts for much of their lives. But sexually mature males must make their own way, and it's a rude shock when their mothers first reject them. Without the guidance of an experienced matriarch, food is harder to find, and young males lose weight and strength. Some do not survive the transition, and those who do go through a few very lean years.

Because of a strict ban on international trade in elephant ivory, Fisher has had little luck obtaining samples from modern animals to compare with his mastodon and mammoth tusks. Soon after he found the signature of first independence in male mastodons, however, he received permission to cut open the tusk of an African elephant that had long been in the collection at the University of Michigan museum. The growth lines inside revealed the same pattern he'd seen in mastodons.

In females, tusk growth rate varies in a different pattern, one that reflects the added need for calcium and phosphate during pregnancy and calving. For both sexes, tusks offer a clear record of the age at which an animal becomes sexually mature. Over the course of 2,000 years at the end of the Pleistocene, Fisher found that the age of puberty in mastodons from the Great Lakes region decreased by about three years, from thirteen to ten years of age. This is the opposite of what you would expect to find if climate change—and a subsequent loss of food sources—was the main stress facing the megamammals.[2] In hungry times, living elephants will delay sexual maturation until the age of twenty or so, building their reserves before they begin the energy-intensive process of mating.

A rush to sexual readiness, on the other hand, fits with the idea that the mastodons were suffering from hunting pressure. Under heavy predation, many animals mature earlier, giving them more of a chance to reproduce before they are hunted down. Examples have been documented in every kind of creature, from fish to large mammals.

Even before he had cracked the tusks' secret code, Fisher was finding evidence that ancient people living around the Great Lakes had dined on mastodon. At sites in Michigan and Wisconsin, he studied mastodon remains that appeared to have been stored on the bottoms of ponds as a

Figs. 9 and 10   Paleontologist Dan Fisher (left) and the mounted remains of the Owosso mastodon (right). Through painstaking analysis of tusk growth rings, Fisher determined that this female mastodon, one of the last of her kind, lost a series of calves before they were one year old. The pattern of life and death Fisher has read from the tusks of many late Pleistocene mastodons suggests that they suffered from heavy hunting by humans. (Photos by Sharon Levy.)

way of refrigerating the meat. The body parts most often stashed in this way were the head and feet, which contain rich portions of fat in the brain and nasal mucosa and in a heavy pad of adipose tissue that cushioned the toe bones from the impact of carrying great weight. More recently, Inuit people have been known to store the nutritious heads of caribou under cold water in the same way.

Fisher speculates that early Americans filled mastodon intestines with rock and sand to anchor butchered body parts. They'd cut a couple feet of intestine, tie an overhand knot in one end, fill it with sand and cobble, and then tie the anchor to the meat masses with a piece of rawhide. He admits that this scenario is based on inference, but the evidence is powerfully suggestive. At two sites, he has been able to culture intestinal bacteria from sediments surrounding fossils found on ancient pond bottoms. When the innards loaded with rock went into the pond, they held microbes native to normal mammalian guts—and these survived for 10,000 years.

Using deer heads and parts of a draft horse carcass, Fisher has tested his theory of underwater meat refrigeration in a local pond. Fresh meat stays on the bottom, in his experience, but long-term storage presents problems. The carcass of an animal killed in fall can be preserved under ice through the winter. In spring, however, the meat is colonized by lactobacilli, which release carbon dioxide gas. The meat mass puffs up, and the slightest breeze can blow it ashore, to be devoured by passing carnivores. A solution is to tether the meat to the pond floor. Then, if you were not a squeamish eater—and Pleistocene people were likely anything but—you retrieved the meat, cut off the moldy parts and ate the good stuff underneath.

Many archaeologists dismiss Fisher's midwestern mastodon butchery sites, because no stone tools remain. He argues that once the animals were dead, people used bone tools to help pry the body apart. Tusk analysis shows that most of the butchered animals died in autumn or winter, setting them apart from another major group of remains he has studied, of males that died battling over mates, always in spring. The timing suggests that the butchered mastodons were actively hunted, not scavenged by people who had found a dead or dying animal.

Fisher keeps his most spectacular example of mastodon butchery, the Pleasant Lake mastodon, in the basement of the University of Michigan museum. The dome of the skull is perforated with multiple holes and score marks that had to have been made by tool-wielding human hands: they look nothing like carnivore tooth marks. At the knee end of a gigantic femur is a clear, wedge-shaped mark made when someone used a piece of bone to lever the lower leg away. The thickest part of the humerus, where the front leg joined the shoulder, had been broken at precise angles, like a piece of worked chert. "You get this result," says Fisher, "not when a wolf or a bear chews the bone, but when a human strikes it with a cobble." The ancient hunters had created sharp butchering tools from the bones of their victim.

There are only fourteen recognized elephant kill sites in all of North America, twelve of mammoth and two of mastodon. That count includes Fisher's Pleasant Lake specimen. To Gary Haynes, an anthropologist at the University of Nevada–Reno, those numbers are telling. Although some researchers claim that so few sites could not possibly indict humans for overhunting, Haynes interprets the figures differently. "If you look at the archaeological record anywhere in the world, there's almost no kill sites of anything except mammoths," he says. "In my mind, the fact that there's twelve in such a short time means there must have been many more we just haven't found."

Haynes, like Fisher, is a self-made expert in Ice Age forensics. The two men have, however, approached the case of the Pleistocene mass extinctions in very different ways. While Fisher began with ancient remains, Haynes focused on newly deceased modern wildlife. First studying the corpses of white-tailed deer in Minnesota, Haynes later moved on to dead moose and wood bison in Michigan and Canada. He searched for carcasses and watched what happened to the bodies in the days and weeks after the animals died. He became intimate with the tooth marks that wolf, coyote, and fox left on bone and compared them to the scrapes left by stone and steel butchering knives.

Then, in 1982, he got a lucky break, a chance to travel to Hwange National Park in Zimbabwe to watch elephants die and decompose. Haynes was not morbidly obsessed—he wanted to learn to read bones accurately in order to understand what happened to the Pleistocene giants. "Most archaeologists, geologists, paleontologists have no idea how the real world operates," explains Haynes. "They have this imaginary past that they dream of, and that has included, and still includes, me."

Haynes arrived in Zimbabwe in a time of intense drought. Hwange's elephants, like the people who worked communal farms nearby, were starving. Haynes had been invited to study the organized killing of elephant herds, done by park managers. Once the animals were shot, the meat was distributed to local farmers and the hides and ivory were sold. The bones left behind allowed Haynes to experiment with stone tool marks and the effects of scavengers and weathering.

Hwange National Park is vast, and in an area where no elephant culling had been done, hundreds of elephants clustered around water holes, slowly dying of starvation. Haynes undertook a comparative study: what different clues remain when a group of animals is killed by harsh climate versus hunting? He used his findings from Hwange to reexamine fossil mammoth sites in Europe and America, where stone tools had been found and which had long been assumed to be the result of mass kills by human hunters.

What he found surprised him. The age profiles of the dead animals at the fossil sites, and the way the bones were distributed, recalled the water holes where elephants in Hwange had died naturally, of hunger and thirst. This kind of death assemblage contains many young elephants and few adult males. Grown males outcompete other animals for food and can walk farther to reach water, so they tend to survive drought better than youngsters do. Haynes also discovered that spiral fractures of long bones, which archaeologists had believed were left by humans extracting the nutritious marrow, are often caused by desperate elephants, who trample the bodies of the dead as they struggle to reach water.

Haynes's work changed the way he and his colleagues looked at ancient elephant remains and undermined the idea that Clovis hunters had slaughtered whole families of mammoths in one go. "Even the presence of stone tools, which people undoubtedly made, doesn't necessarily mean that people killed the animals," he says. "In nature, elephants die en masse sometimes. You don't need people to make that happen."

Nevertheless, Haynes thinks human hunting played an important role in the demise of America's native elephants. Mammoths and mastodons would have traveled along obvious trails, lined with dung, as African elephants do today, and would have limited their movements to places within a day's walk of drinkable water. As a warming climate altered late Pleistocene vegetation, the beasts would have been drawn to small patches of remaining habitat, making them easy for Clovis hunters to find.

The ability to decipher the secrets of mammoth dung was likely among the first skills that people developed as they moved into the New World. By the dawn of Clovis time, the woolly mammoths of the far north were already rare, but Columbian mammoths remained common in the continental United States. Taller than their woolly cousins, the mammoths of the lower forty-eight states would have been capable of traveling far and fast, more than 20 miles a day, often passing bits of unchewed plants, sand, or soil far from their last feeding ground. From this evidence, an observant hunter would be able to deduce an animal's feeding range. He could look at a bolus of elephant dung and estimate not only the size and age of the animal that left it, but how fast the beast was moving and whether it was feeling well. All this can be read from the size and texture of the feces, and how much it splattered when it hit the ground.[3]

In Haynes's model, climate change made the megamammals vulnerable and humans gave them a killing blow, using the mammoths' own well-worn trails as a guide. He admits there is no solid archaeological evidence of Clovis tracking behavior. He's reaching back into Clovis time in his imagination, using the more than two decades of experience of elephants' spoor, of their lives and deaths, that he has gained in southern Africa.

Before a twentieth-century boom in the international ivory market led to a hunting frenzy that devastated herds across Africa, elephants walked traditional migration routes that followed the easiest contours across hilly terrain. These trails could stretch for hundreds of miles. An elephant route in Uganda was at one time regarded as the best road in the whole country.[4] America was once similarly veined with mammoth trails that took the most efficient cross-country routes. Many of them now lie beneath our highways, says Haynes, so he admits that it is unlikely anyone will ever find ancient mammoth paths lined with Clovis points. Still, when he tries to imagine Pleistocene North America, he pictures the landscape of Hwange National Park, marked with obvious elephant routes.

A rare, tantalizing bit of evidence supporting this vision was uncovered in 1998, when the St. Mary Reservoir in Alberta, Canada, was drained to allow construction of a new spillway. When the water ran off, it revealed a wide expanse of ancient riverbed, containing tracks and

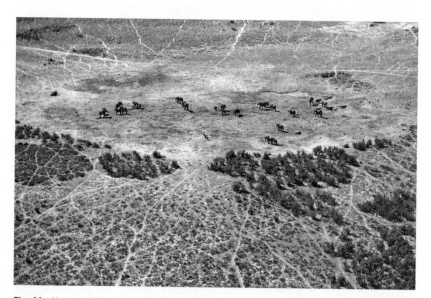

Fig. 11   Hwange National Park, Zimbabwe, is crisscrossed with elephant trails. Anthropologist Gary Haynes believes that late Pleistocene North America was likewise marked with obvious trails that led Clovis hunters to the haunts of mammoths and mastodons. (Photo courtesy of Gary Haynes.)

bones of extinct horses, musk oxen, camels, and mammoths. The bones all dated to about 13,000 years bp. Also found were Clovis spearpoints that held traces of horse blood, indicating that human hunters had once walked among this Pleistocene menagerie. Paul McNeil, then a graduate student studying dinosaur locomotion at the University of Calgary, was called out to the site, known as Wally's Beach, to analyze the tracks.

Few preserved mammoth trackways have ever been found, and for McNeil, the chance to read their movements written in the Alberta sand brought the giants rumbling back to life. He had been skeptical when he first heard about the discovery but was convinced the moment he saw the tracks: large circular impressions, 20–24 inches across, with a stride length of 6.5 feet. The only creature that leaves those kinds of prints is an elephant, and in Alberta, that meant woolly mammoth—the larger Columbian mammoth probably never lived that far north. Woolly mammoths were about the same height as modern elephants, so McNeil used data on African animals to interpret the size and age of the animals that left tracks, as well as the condition of family groups.

In places, the wide tracks of a mother ran side by side with the small prints of a very young calf. It was clear from the stride lengths that the adult had been walking slowly, while the juvenile was almost running to keep up. Another tract of sand recorded a moment of play between two juvenile mammoths, one a teenager and the other about ten years old. As they chased each other back and forth, the youngsters left arcing tracks, an enduring echo of their ancient romp.

The traces preserved at Wally's Beach show that Pleistocene horses, camels, mammoths, and musk oxen coexisted with people for about 300 years on the banks of the St. Mary River. Sets of tracks averaged over that time span suggest a population of mammoths dominated by adults, with only about 30 percent juveniles.[5]

According to Haynes, a healthy herd of modern elephants should be about 50 percent juveniles. A small proportion of young animals in a herd means that something is wrong. The calves may be dying off because they make easier targets for predators, including hunters with spears. They are also less likely than adults to survive a drought. In the late Pleistocene, as the glaciers sent meltwater cascading out of the north, Wally's Beach was unlikely to suffer from drought. The tracks there offer a glimpse of the lives of one mammoth population, in a small fraction of their range, and are far from conclusive. But they fit with the idea that human hunters made a serious impact.

To understand how hunters armed only with Stone Age technology could drive mammoths and mastodons into oblivion, it's important to understand the ponderous process of elephant reproduction. In the wild, modern African elephants bear their first calf at about age twelve. In good times, when enough food and water is available, they will birth a calf every four years. In times of drought, most females won't go into estrus at all.

Norman Owen-Smith, an ecologist at the University of Witwatersrand in South Africa, has made a careful study of what he calls megaherbivores: elephants, rhinos, and hippos, the only extant land mammals that weigh more than 1,000 kg (approximately 2,205 pounds) as adults. Their size, he has found, makes these animals ecologically unique. Adults are so massive that they are immune to the attacks of lions and other large predators, yet megaherbivores are particularly vulnerable to the impacts of human hunting.[6] Owen-Smith calculates that the potential growth rate of an African elephant population is no more than 6.5 percent per year. Such a slow-breeding species, he says, would be doomed if any efficient predator began to focus on it. This, he believes, exactly describes the situation of Pleistocene elephants in the Americas. Owen-Smith interprets the fact that the great majority of known Ice Age kill sites involved mammoths or mastodons as evidence that they were humans' preferred prey. There were plenty of smaller animals around to sustain them if they couldn't bag a megamammal, so human hunting pressure did not let up as elephant numbers dwindled.

The calculus is very different for some of the less formidable herbivores that vanished at the end of the Pleistocene—native American horses, for instance, and the stag moose, an outsized deer. These creatures bred much faster and were likely more numerous than mammoths and mastodons. In theory, they should have been able to withstand hunting pressure, as the bison and the white-tailed deer did. Owen-Smith argues that the elephants went first, and with their passing, entire landscapes changed forever.

Elephants break major branches and sometimes knock over whole trees when they feed, opening up woodlands and transforming them into savanna or shrubland. Where elephants live, habitats become open enough to support grass, and there is soon enough grass to support widespread fire, another strong force for vegetation change. These elephant-driven habitat characteristics benefit a range of other herbivores in Africa. Recent research shows that this effect applies not only to large mammals such as zebra, but also to rodents and geckos that shelter in convenient crevices in elephant-mangled acacia trees.[7]

As Owen-Smith sees it, midsize herbivores in Pleistocene America died out from a combination of human hunting and unfavorable habitat changes that followed from the demise of megaherbivores. With these monster vegetarians gone, there was no effective force to counter the thick growth of forest that accompanied the warming weather at the end of the Ice Age. Animals that had survived earlier interglacials by moving north were trapped by impassable woodland. The loss of mammoth and mastodon affected every living thing: lizards, horses, trees, even uncropped tufts of grass.

For Fisher, counting up hunted elephants is no longer important. No matter how many kill sites are found and documented, they will always represent a miniscule fraction of Pleistocene mammoth and mastodon populations. The preservation and rediscovery of killed animals require a complex series of natural accidents. In his view, it will never be possible to show, based on the number of such finds, that humans drove the extinctions. "Show me the evidence that elephants were responding to human predation, or to climate change, as an ecological force," says Fisher. "I think that evidence is going to come from tusks."

There is no sign in the tusk record that either mastodons or mammoths were going hungry late in the Pleistocene—the pattern that would corroborate the climate change theory of extinctions. Annual growth increments vary year to year for any individual but show no pattern of growing thinner as the Ice Age waned. Analyses of the nitrogen isotope content of late Pleistocene mastodon tusks from the Hiscock site in western New York suggest the animals were, on the contrary, robust and well fed.[8] Oxygen isotope analysis, which reflects the shifting of temperatures with the seasons, suggests that Siberian mammoths were able to thrive in a wide range of weather conditions, including those that prevailed at the time the last mainland mammoths died out.[9]

If people had never colonized North America, says Fisher, the continent would still be home to mammoth and mastodon. His detective work reveals the ways of extinct elephants in the kind of intimate detail that can be difficult to track even in living species. He has begun to analyze a record of musth, the elephant version of rut, recorded in male tusks as a seasonal thinning of the dentin layers that begins after twenty years of age. Male elephants leave their natal groups in their early teens, but most do not mate until they go into musth—a heightened hormonal state that can turn otherwise calm animals into aggressive hotheads. Musth is demanding: in its grip, males fast and expend great effort in pursuing females in estrus and fighting other males for access. After up to three months of this intense action, males grow thin and tired, their musth stops, and they return to their normal haunts, away from female family groups, to eat and restore their strength in peace.

Based on his study of bull mammoths and mastodons, Fisher describes "bone-smashing, tusk-cracking, hide-rending musth battles" that took place across North America. The skull of the Hyde Park male and several others show the resulting injuries, obvious even to a novice's eye: gaping holes in one or both cheek bones, caused by the deadly upward thrust of an opponent's tusk as two bulls clashed head to head. From the Hyde Park tusk, Fisher has teased out evidence that the big male administered many such blows before dying of one himself. Stabbing upward at his enemies rocked the mastodon's tusk in its socket, damaging the soft tissue that secreted ivory and causing a warp in the pattern of tusk growth.

Fisher is beginning to suspect that late Pleistocene mastodons were going into musth at unusually young ages—a sign that they lived in a declining population whose social structure was disintegrating. An incident from the mid-1990s in Pilanesberg National Park, South Africa, provides a modern comparison. Elephants in the Pilanesberg area had been wiped out by hunting. When the park was established, young orphans from Kruger National Park were brought in to reestablish an elephant population. Kruger had been deemed overpopulated, and park managers "culled" some family groups, methodically killing every adult. In normal elephant societies, mature bulls keep the younger males in line and regulate the onset and length of the youngsters' musth. Elephants seem to need time to learn to cope with the heavy rush of testosterone and stress that musth brings—in the stable, wild population at Kenya's Amboseli National Park, musth lasts only a few days in younger bulls but increases with age, lasting as long as three months in the eldest, most experienced animals. There were no mature bulls in Pilanesberg, and in their absence, several young males went into musth at younger ages and for longer periods of time than had previously been known. Hyped up on hormones, the young bulls ran amok, attacking and killing endangered rhinos and, in two cases, people. Pilanesberg's angry young elephants were brought under control when several older males were transplanted from Kruger National Park. The males each found a niche in a hierarchy ruled by the older bulls, their periods of musth grew shorter and less frequent, and the unusual attacks stopped.[10]

For Fisher, the Pilanesberg story raises intriguing questions about the future of modern elephants, as well as the last days of the mastodon. How long do you have once you start radically reducing population numbers, he asks, before animal populations start to self-destruct? As he delves deeper into the record left in tusk diaries and other revealing fossils, Fisher is uncovering more clues to the turbulent lives of the last American elephants. Some female mastodon skulls from very late in the Pleistocene show signs of the kind of severe battering normally seen only among musth bulls. From his detailed studies of butchered mastodon remains, Fisher has concluded that the earliest human hunters in the Great Lakes region targeted solitary males. Perhaps this new threat was enough to wipe out the oldest and most experienced bulls, leaving an unruly mob of young males. These unlucky cows may have been beaten up in conflicts with out-of-control young bulls like the ones at Pilanesberg.

The biographies Fisher reads from tusks make static old bones come alive in a way no museum exhibit can. The most poignant story he tells is that of a female mastodon from Owosso, Michigan, north of Ann Arbor. She broke one of her tusks in half during a life punctuated with tragedy: she lost a series of calves, one after the other, a sign that in her time—and preliminary carbon-dating suggests she was one of the last few of her kind to walk the Earth—mortality rates exceeded even those seen in some poached herds of African elephants today. Owosso died young, in her twenties. After her death, someone cut off her fat-rich toes—so whether they killed her or stumbled across her carcass, humans ate parts of her body. Today, her skeleton stands wired together in a high-ceilinged hall at the University of Michigan museum. Her fleshless skull and empty eye sockets evoke pity for the hardships she suffered so long ago. Fisher believes it is important to tell this long-gone mastodon's story. Only by looking to the past, he says, can we understand the consequences of our actions in the present. Owosso's sad tale offers a way to demonstrate the powerful human influence on the future of our planet, for good or ill.

Many scientists now see the dilemma of endangered African and Asian elephants, trapped in shrinking remnants of habitat, as a mirror image of the last days of the mammoth and mastodon. Surviving pachyderms offer clues about how and why their lost cousins perished. Ancient bones may help teach us how to keep the remaining giants alive.

OF MAMMOTHS AND MEN

The clues left behind in the bones and tracks of vanished mammoths suggest they lived in societies very much like those of their modern African and Asian cousins. At Amboseli National Park, Cynthia Moss, Karen McComb, and their colleagues have followed the lives of more than 1,700 individual elephants over the course of more than thirty years.[1] The troubles of these living elephants can fill gaps in our knowledge of America's extinct giants, as well as informing urgent conservation measures today. The normal life span of an elephant—or a mammoth—is the same as a human's, about sixty-five to seventy years. If an animal reaches adulthood, avoiding the most common disasters—poaching or death by starvation in times of drought—she is likely to live until her massive molar teeth are worn to the nub and she can no longer chew her food.

The Amboseli researchers have found that families headed by older matriarchs are more successful in raising calves than those with younger leaders.[2] Older females know the best ways to find food and avoid conflicts with people. Careful study of the way elephant families react to encounters with others of their kind also shows that older matriarchs are more knowledgeable about the structure of the wider elephant society they live in. They recognize the unique contact calls of hundreds of other females and understand which animals can be trusted and which must be defended against.

The eldest are, alas, the most valuable not only to elephant families but also to human poachers. Among African elephants, both males and females grow tusks, and the older the animal, the larger and therefore more valuable the tusk. So the oldest and wisest elephants are most likely to be targeted by poachers looking to profit from a booming black market in ivory. "If groups rely on older members for their store of social knowledge," write McComb and her coworkers, "then whole populations may be affected by the removal of a few key individuals."[3]

In recent years, conflicts between African elephants and the people living around them have escalated. The number of people killed in elephant encounters is rising. Some researchers believe that wild elephants are becoming more aggressive—toward people and toward one another—because of long-term impacts of the trauma they suffer during poaching and legal culling. In healthy populations, young elephants grow up constantly attended not only by their own mothers but by a circle of doting aunts, cousins, and grandmothers. Like human children, elephants can become seriously maladjusted when they are orphaned by the violent removal of adult family members.

Ecologist and psychologist Gay Bradshaw and colleagues summed up the situation in an essay published in *Nature* in 2005: "Elephant society in Africa has been decimated by mass deaths and social breakdown from poaching, culls and habitat loss," she wrote. "From an estimated ten million elephants in the early 1900s, there are only half a million left today. Wild elephants are displaying symptoms associated with human Post-Traumatic Stress Disorder: abnormal startle response, depression, unpredictable asocial behavior and hyperaggression."[4]

The rhino-killing male elephants at Pilanesberg, she explained, had not only suffered the trauma of seeing their mothers and aunts killed. They had also missed out on normal social interactions with female relatives as juveniles, as well as the time they ought to have spent with older males after they left their natal herds as young adults. The intense aggression the Pilanesberg elephants displayed is apparent in other dwindling, isolated populations. At one African park, 90 percent of all male deaths are caused by fights with other bull elephants.

Eve Abe, a Cambridge-trained ethologist and member of the Acholi tribe from northern Uganda, has studied the psychological parallels between orphaned elephants in Queen Elizabeth National Park and Acholi children forced to fight in the long war between the Ugandan government and the Lord's Resistance Army. The LRA often murdered parents before abducting their children, and elephant poaching intensifies in war time—smuggling ivory is an easy way to finance an army. So in Uganda, elephants and people have suffered in very similar ways.

"We used to have villages," Abe told a *New York Times* reporter in 2006. "Everyone in northern Uganda now lives in displaced people's camps. The elders were systematically eliminated. All these kids who have grown up with their parents killed—no fathers, no mothers, only children looking after them. They form these roaming violent, destructive bands. It's the same thing that happens with elephants. Just like the male war orphans, they are wild, completely lost."[5]

If our Pleistocene ancestors did help destroy the last of the mammoths, part of the cause, and part of the tragedy, may have been that people and elephants are so alike. Both humans and elephants evolved in Africa, and spread from there to the far-flung reaches of the planet. The living elephants of Asia and Africa, along with the now-vanished mammoth lineage, descended from quadruple-tusked African ancestors six million years ago. The oldest known evidence of hominids dining on elephant comes from Olduvai Gorge in Tanzania and dates to 1.8 million years ago. (The hunters were probably *Homo erectus*, the first of our ancestors to get seriously into meat-eating.) That kill was made at the dawn of the Pleistocene, the epoch that would see the rise of modern humans and a near-complete loss of pachyderms worldwide.

In Pleistocene times, elephants or their relatives lived on every continent except Australia and Antarctica. Mastodonts, shorter, stockier cousins of the mammoth, roamed from the southern tip of South America to the northeast of what is now the United States. The Columbian mammoth reigned throughout western and southern North America, and woolly mammoths flourished in a cold, arid grassland that stretched across northern Europe and Asia and into Alaska and Canada.

Todd Surovell, an archaeologist at the University of Wyoming, argues that elephant extinctions followed the movement of modern humans out of their homeland in Africa. The gradual northward range contraction of Old World elephants, he says, followed by their ultimate extinction in Europe and the Americas, traces the path of hominid population expansion.

Surovell analyzed known sites worldwide containing elephant bones along with strong evidence that hominids had killed or butchered the animals. Demonstrating that a fossil elephant was killed by people is difficult.

With smaller creatures, elk, for example, hunters cut off body parts to transport them back to camp. Elephant bones and meat, on the other hand, are so heavy that people instead moved their camps to the carcass. Sites widely accepted by archaeologists as examples of human exploitation of elephants show a pattern of one or a few dead animals, surrounded by flakes of chipped stone created as the hunters worked their butchering tools to a sharp edge. In a few cases—most often at North American sites—a stone spearpoint was left buried among the bones.

By 500,000 years bp, modern *Homo sapiens* were hunting elephants in the Mediterranean and left a trail of corpses behind as they moved north through Europe and Asia. About 120,000 years ago at Lehringen, Germany, hunters killed a woolly mammoth using only wooden sticks whittled to a sharp point and hardened by the heat of a camp fire; by some miracle of preservation, one of these ancient weapons was found among the giant's ribs.

As glaciers retreated at the end of the Ice Age, our ancestors penetrated the high Arctic, and woolly mammoths died out except in a few remote refuges bypassed by the human tide. Soon after, Clovis culture spread quickly across North America, and mastodons and Columbian mammoths vanished. Among the last elephant populations to blink out were gomphotheres, members of the mastodon clan found in Central and South America.

"We know from studies of modern hunter-gatherers and from many archaeological studies that humans tend to target the largest available species," explains Surovell. "There's good reason for this. You get much greater returns relative to the time and energy invested in the hunt, more bang for your buck." Some anthropologists believe prehistoric elephant hunting was also driven by a desire for social prestige. Bringing down a mammoth was a way for a man to get plentiful food to share, and therefore great respect.

For Surovell, the most significant evidence for human-caused extinction is the staggered timing of elephant disappearances worldwide, which align with no single climate event. "The pattern of elephant extinction in space and time cannot be explained by climate change," he says, "but it perfectly tracks human colonization around the globe."

The picture Surovell paints is persuasive. However, it does not agree with all the emerging evidence on life during the Pleistocene—especially in the far north. That region holds a rich record of both people and woolly mammoths, the member of the elephant family at home on the frigid prairies that lay just south of the ice sheets.

In Europe and Asia, Pleistocene men, women, and children left behind cave paintings, engravings, and sculptures that record their deep knowledge of the animals around them: woolly mammoths, wild horses, ibex, reindeer, cave lions, and aurochs, the progenitor of domestic cattle. At Rouffignac Cave in France, someone drew the outlines of two hulking male mammoths facing off, in what could be read as either a greeting or a prelude to battle. Others etched mammoth images into pieces of stone, bone, or antler. Some made spear throwers in the shape of a mammoth, using the space below the carved tusks to hold a spear shaft in the instant before it flew.[6]

"Here are people living for tens of thousands of years around these animals," says Dale Guthrie, a paleobiologist at University of Alaska–Fairbanks and long-time student of Pleistocene art. "There are very few instances of woolly rhino or mammoth drawn with spears sticking out of them. With horse, bison, ibex, reindeer, there are lots of images of those animals speared, blood coming from the mouth. You might judge from that, that the biggest, most dangerous animals weren't hunted too much."

In fact, the archaeological evidence from northern Europe and Asia shows that Pleistocene people there hunted mostly reindeer, elk, and other members of the deer family. In some cases they seem to have tracked the course of reindeer migrations, intercepting the animals at river crossings where they were most vulnerable. They did sometimes pursue mammoths, but mammoth-hunting was not their mainstay.

The modern humans who became the first to survive in the forbidding landscape of Ice Age Siberia succeeded not because they were expert elephant killers but because they learned to take what they could get. The occupation sites they left behind are cluttered with bones of snowshoe hare and geese. Living in bodies designed for tropical heat, these ancient people conquered extremely cold environments through sheer inventiveness. They learned to sew warm clothes, to capture fish, birds, and small mammals, and to build underground shelters.

Guthrie has spent decades exploring the vanished world of the mammoth. A bow hunter, he also stalks the wildlife of present-day Alaska. On a recent trip to the Bering Sea Islands, he and his son put in 40 miles of patient tracking in pursuit of a bull musk ox. A rare relic from the days of the mammoth, Alaska's small population of musk oxen is meticulously monitored by the state's Department of Fish and Game. "As a hunter, I know what humans can do if they're unregulated," says Guthrie. "Hunters can overkill, and I used to believe that was what happened at the end of the Ice Age." Now, he believes the truth is more complex. He has published a series of papers showing that megafauna in Alaska and the Yukon were hard-hit by climate change even before any people arrived.

When Guthrie took a professorship at the University of Alaska–Fairbanks in the 1960s, he had been taught that Ice Age ecosystems were the same ones that exist today and that prairie or boreal forest simply migrated north or south as the glaciers retreated or expanded. He was among the first of a new wave of researchers to argue that the landscape of the Pleistocene was a foreign one, filled with strange communities of plants unlike any recorded in historic times. The evidence for this came not only from old bones but also from analysis of insects, plant parts, and pollen preserved in Ice Age soils, which contained a bewildering jumble of organisms that now live nowhere near one another.

The abundance of fossil mammals then being unearthed in Alaska—lions, woolly mammoths, horses, camels, reindeer, short-faced bears—had coexisted on an icy prairie, an ecosystem Guthrie dubbed the "mammoth steppe." Few trees grew there—it was too cold, windy, and dry, and the more the glaciers expanded, the more arid the steppe became. The giant creatures of the mammoth steppe had lived under "cloudless skies that were brilliant in summer and blue frigid in winter"—a world radically different from the soggy, cloud-covered tundra that lay outside Guthrie's house in Fairbanks.[7]

When the mammoth steppe faded away at the end of the Ice Age, many of its denizens, like the mammoth, woolly rhino, and short-faced bear, went extinct. But others survived in dramatically shrunken ranges, such as lions in Africa and Przewalski's horses and saiga antelope on remnants of steppe habitat in central Asia. Meanwhile, a new set of megafauna—moose, elk, and bison—invaded the north.

To understand the varied fates of these big herbivores, Guthrie looks to their innards. Any creature that lives off leaves or stems must crack a formidable array of defenses to absorb the nutrients inside. Many plants produce toxins designed to discourage herbivores—and the longer the vegetation grows, the more poisonous it gets. Even the youngest, freshest buds and leaves are full of cellulose, the complex sugar that builds plants' cell walls. Only bacteria are able to break down cellulose, so every herbivore from rabbit to elephant has evolved ways to nurture cellulose-digesting bacteria in its gut.

Wild ruminants like bison and moose ferment their food in a large forestomach called the rumen. Inside this chamber, symbiotic microbes tear apart plant cells and reconstruct their contents into a complete nutritional package that includes every B vitamin and every essential amino acid. Because they ferment their food before it reaches their intestines, ruminants can eat toxic plants that would sicken or kill a horse. However, their digestion is slow by comparison—and because of the way their guts are designed, it cannot be speeded up.

Elephants and horses also ferment their food, not in a rumen but in a large pouch called the cecum, which lies between the small intestine and the colon. The fermentation process is the same in both groups of animals, yet their survival strategies are very different. Since a horse's fermentation chamber lies farther along in its digestive tract, it can absorb proteins that would otherwise be broken down by microbes. Horses can make do on tough, old grasses—a diet that could never sustain a ruminant. If the available forage is poor, heavy on useless fiber and low on nutrients, horses respond by eating constantly, running more and more food through their systems. The dry, cold plains of the Pleistocene had a short growing season but lots of grass. An ability to make it through long winters on tough, withered stems gave hindgut fermenters, like horse and mammoth, a major advantage.

As the Ice Age faded, willows began to colonize the banks of newly forming streams. Other trees and shrubs followed; many of them heavily armed with toxins only a ruminant could digest. In Guthrie's view, these habitat shifts drove the decline of the horse and mammoth.

Guthrie has gathered an impressive collection of old and new radiocarbon dates on fossil mammoth, horse, bison, elk, and moose from Alaska and the Yukon. The patterns of animal distribution in time and space show that the transition from dry steppe to wet tundra kicked in between 14,000 and 15,000 years bp. The last of the region's horse fossils date from that window of time, a moment when bison, elk, and moose— all ruminants—exploded in numbers and moved far into the north.

Mammoths seem to have become less common around the same time, and vanished by 13,000 years bp.[8]

To further support his climate scenario, Guthrie has shown that Alaskan horses began shrinking thousands of years before the first people settled the region, continuing to grow steadily smaller until they disappeared.[9] That kind of pattern implies a gradual response to changing climate, rather than a sudden impact of human hunters. Guthrie bases his claim on measurements of hundreds of Pleistocene horse metacarpals, foot bones that are critical in carrying the animal's weight and thus make a good indicator of its body mass.

The data show that woolly mammoths and humans coexisted in Alaska for at least a few hundred years, and Guthrie agrees that people may have helped drive the mammoth into oblivion. If this happened, however, he emphasizes that the giants' condition had already been weakened, the result of a climate change that degraded the hindgut heaven where they had roamed for so long.

The occupation of Siberia that would lead to the peopling of the Americas started 15,000 years ago, when the glaciers receded, uncovering long-frozen land. The mammoth steppe cracked open, split with newborn streams that carried meltwater from the glaciers. Mammoth populations began to dwindle as the animals became isolated in remnant patches of steppe habitat. People appeared in growing numbers as they followed elk and moose into the far north.[10]

Woolly mammoths held on longest in the Siberian Arctic, in a few remnant pockets of arid steppe habitat. The population that survived longest was on Wrangell Island off the northeastern coast of Siberia, where a few mammoths lived until 4,000 years bp, finally dying out when the first humans reached the island.[11] These last woolly mammoths eked out an existence on a postage stamp of tundra steppe off the Siberian coast—while on the other side of the planet, the Egyptian pharaohs built their pyramids.

The woolly mammoth's demise was thus slow and gradual. Humans likely gave the species a killing blow, but this was possible only because climate change had already wiped away most of their habitat. Had humans left Beringia unexplored, small populations of mammoth might well have lived on, as they had during previous warm spells.

In a pattern that echoes the last days of the woolly mammoth, elephant populations in Africa and Asia today are being isolated in small remnants of habitat. In an elegant but grim study of human and elephant communities in West Africa, Richard Barnes of the University of California–San Diego demonstrated that a severe drop in elephant numbers matched perfectly with a human population boom during the twentieth century.[12] Intense ivory hunting during colonial times, between 1814 and 1914, had killed off about 75 percent of the region's elephants. Even if the animals had not died then, Barnes argues, they would surely have suffered later when the rural human population grew and took over resources both species need to survive: fertile soils and water. In what are some of the poorest nations on earth, elephants raiding the crops of subsistence farmers cause terrible hardship. For people living at the edge of shrinking elephant reserves, there is no incentive to protect the giants.

When elephants are isolated in a growing sea of humanity, long-term survival becomes a numbers game. In the 1980s, a spike in ivory prices led to an elephant-hunting binge throughout the animals' range. A study in Ivory Coast showed that the odds of an elephant population surviving the decade depended on the amount of habitat available. Groups in ranges of less than 500 square kilometers (193 square miles) had a 50 percent chance of making it; those with at least 750 square kilometers (about 290 square miles) in which to roam endured.[13] But in the ongoing struggle for land, pachyderms continue to lose ground to people.

Adrian Lister, a mammoth expert at University College London, also studies modern Asian elephants in Borneo. He sees the predicament of the highly endangered elephants as very similar to the one that doomed their Ice Age cousins: a combination of massive habitat loss and heavy human hunting. Forests are rapidly being eliminated from the Asian elephant's range for agriculture and logging, he explains, so the animals find themselves restricted to small pockets of habitat. "If you map the present distribution of the Asian elephant, it looks like an outbreak of measles: lots of small, isolated populations all over southeast Asia, where formerly it was a continuous swathe from China almost to the Mediterranean." Asian elephants thrive in forests, while mammoths preferred grassland, but the principle is the same. Only the cause of habitat loss is different: climate change in the Pleistocene, the impact of a growing human population now.

Meanwhile, in both Asia and Africa, hunting pressure continues. Despite a 1990 ban on international trade in elephant ivory imposed by the Convention on International Trade in Endangered Species (CITES), poaching remains big business. The impacts of this black market industry are difficult to track. Still, the number of tusks seized by law enforcement agents in 2005 suggests that about 23,000 elephants were slaughtered for their tusks in a 12-month period.[14] In recent years, increases in the ivory trade have corresponded with major decreases in elephant numbers, even in areas where habitat loss, the other driver of elephant decline, has remained unchanged.

In the 1980s, prior to the ban, about 100,000 animals were being killed per year, and 80 percent of herds were lost in some regions of Africa. It's an old story—in the 1800s, a booming demand for ivory in Europe and the United States led to a binge in elephant hunting in Africa. By the turn of the twentieth century, many elephant populations were completely wiped out, and others were reduced to a handful of survivors.

In a few corners of southern Africa, elephant numbers have rebounded. Today, 70 percent of Africa's elephants live in the southern part of the continent, in reserves and national parks that cover a small fraction of their former range. In these places elephant management is a hotly contested issue, and the need to understand elephants' past is urgent.

Kruger National Park in South Africa and Chobe National Park in Botswana are two prime examples. In both places, authorities designated protected lands in the early 1900s, a time when the elephant population had been decimated by hunting, and rinderpest, a virulent viral infection, had knocked down numbers of other large browsing animals. The acacia trees that would previously have been broken off and trimmed by hungry elephants had grown tall and stately.

Kruger contained no elephants when it was declared a reserve, but once the land was protected elephants crossed over the border from Mozambique and began to multiply. By the 1960s, the megaherbivores were having a visible impact on their habitat—particularly on the acacias, which were reverting to their normal shrubby state. At this point park managers decided to kill off enough animals to keep the population below 7,000, the figure they deemed to be the park's "carrying capacity" for elephants.

The culling decision was based on concepts from agriculture, not ecology. The idea that elephants and the plants they feed on could, and should, be kept in a static balance recalls management principles applied to ranchlands where domestic cattle are raised. Rudi van Aarde, a zoologist at the University of Pretoria, says that culling targets were "arbitrary, whimsical and had no theoretical basis." Lindsey Gillson, an ecologist at the University of Cape Town, notes that culling became, in practice, a policy of sustained ivory production. The stated goal of maintaining savanna woodlands in their "pristine," turn-of-the-century condition made no sense. "Ecologists and wildlife managers attempted to achieve the impossible goal of maintaining viable elephant populations *amidst a vegetation pattern that would only occur if no elephants were present*," she and Keith Lindsay wrote.[15]

Over the last four decades, ecologists have given up the once-popular idea that ecosystems naturally reach and remain in stable states. The old paradigm of the balance of nature has given way to a new concept, what Gillson calls the "flux of nature." This principle of ever-shifting communities has been studied from the kelp beds of the Pacific Ocean to the African savanna. There is now strong evidence that precolonial savannas never reached an enduring balance, but instead moved through cycles of changing abundance among elephants, trees, and grasses.

There is no reliable way to tell how dense elephant populations were in precolonial Africa. Archaeological and historical evidence shows that they had been hunted for an international ivory trade for many hundreds of years, but the take could not have been nearly as high as it grew in the nineteenth century. In the 1970s, zoologist Graeme Caughley used the history recorded in the twisted trunks of mopane trees in Zambia's Luangwa Valley to build evidence of a long-standing, ever-shifting relationship between elephant density and the extent of mopane woodland. When elephants were abundant, they would knock trees down before they could grow tall. When the mopanes dwindled from this abuse, elephant numbers would decline and in time the trees would reach toward the sky. Then the cycle would begin again.

Elephants break off straight mopane trunks as they browse, causing the stumps to resprout in a forked shape. Caughley's survey of older mopane trees in the Luangwa showed that most had been broken and forked long ago, when they were saplings.[16] This meant that before the first European explorers described the area, elephants were thick on the ground—likely as dense as they are today in some reserves in southern Africa. This evidence was intriguing, but inconclusive. Caughley predicted that if he was right the elephant/tree cycle would repeat about every 200 years—and he invited other researchers to prove him wrong.

At the time of Caughley's study, most ecologists believed that without human interference, elephants and savanna trees would naturally reach a stable balance. The much-discussed "elephant problem"—the fact that in many parks and reserves elephants had begun to kill off savanna trees faster than they could regrow—was assumed to be the result of human land use that kept too many animals trapped inside artificial boundaries.

At the turn of the twenty-first century, Gillson found support for Caughley's ideas in fossil pollen samples collected from Kanderi Swamp in Kenya's Tsavo National Park. Her data show that over the last 1,400 years, the number of savanna trees there rose and fell in cycles lasting about 250 years. Acacia woodland would dominate; then the trees

Fig. 12   These mopane trees in Hwange National Park were broken off by browsing elephants. Today, as in the distant past, elephants shape their habitats by breaking trees and encouraging the growth of grass. The disappearance of mammoths and mastodons may have affected many other Pleistocene creatures adapted to elephant-influenced environments. (Photo courtesy of Gary Haynes.)

would dwindle and grasses would reign for a century or two. The cyclical patterns of tree abundance are consistent with Caughley's prediction of woodland–grassland cycles driven by elephants, but the cycles revealed in the ancient pollen record were likely driven by wildfire as well as browsing pressure.[17] Studies of swamp sediment show a clear correlation between the amount of charcoal left by fire and shifts in pollen abundance, but there is no direct proxy for elephants in the microfossil record. Eventually, paleoecologists working in Africa may be able to use fossil spores of the dung fungus *Sporormiella* to trace the abundance of elephants in the distant past, as Guy Robinson has done for megafauna in the northeastern United States. Gillson is now testing the technique in Hluhluwe Reserve, where she can track the amount of fungus left behind in the soil by a known density of elephants.

A recent study of Chobe National Park concluded that there is no ecological reason to change the numbers of elephants there, though the authors acknowledge that there may be compelling sociopolitical reasons. The disappearance of acacia woodlands along the Chobe River is still seen by many as an "elephant problem," and elephants have indeed reshaped the Chobe waterfront. Ultimately, this appears to have boosted, rather than impaired, biodiversity. Numbers of buffalo and impala have grown along with the elephant population. The diversity of small mammals and birds has increased.[18]

About 50,000 elephants are thought to survive in Asia, and an estimated 400,000–660,000 in Africa. But many areas of potential habitat have not been surveyed for years, if ever, and some populations have likely succumbed to ongoing habitat loss as desperate farmers try to cultivate even unproductive land. No one really knows how many elephants live on. Populations that once roamed throughout the two continents, numbering in the millions, have suffered a wave of extirpations that must resemble those that hit mammoths and mastodons at the close of the Pleistocene.

In both Africa and Asia, growing human populations are converting more and more elephant habitat into cropland. With increasing frequency, elephants are raiding crops at the edge of wildlife reserves, bringing them into a conflict with farmers that can prove deadly for both sides.

To understand what impels elephants to raid crops, Thure Cerling, a geophysicist and biologist at the University of Utah, used stable isotopes of carbon and nitrogen to track eating habits in a herd in Kenya's Samburu National Reserve. Stable isotope ratios have been used to study the diets of fossil animals. (Mastodons, it turns out, browsed on trees and shrubs, while mammoths were grassland feeders.) The technique holds promise as a new tool that can help wildlife managers to keep modern elephants out of dangerous conflicts with people.

Low ratios of carbon-13 to carbon-12 reveal a diet of trees and shrubs, while high ratios indicate consumption of grasses or grain crops such as corn and millet. Ratios of nitrogen-15 to nitrogen-14 show where the animals have been foraging—dry areas like the Samburu have high ratios, while the wetter forests on Mt. Kenya have low ratios. The travels of elephants in the study were also tracked using GPS collars.

Cerling and his colleagues found that most of the elephant herd spent their time in the arid lowlands of the reserve. In the dry season, they ate trees and shrubs; when new growth sprang up in the wet season, they switched to grasses. The lone exception to this pattern was a bull the researchers named Lewis. He spent the rainy seasons in the safe haven of Samburu, but when the weather turned dry he moved 25 miles cross-country to a forest 6,500 feet up on the flank of Mt. Kenya. He sheltered there during the day, making nighttime raids into the corn fields of nearby subsistence farmers.[19]

Corn is more nutritious than forest browse, and Lewis risked conflict with humans to get at the best food, following his innate drive to stay strong for future musth battles with other males. "It's a high-risk, high-gain strategy," says Iain Douglas-Hamilton, a coauthor of the study and founder of the Save the Elephants foundation. "In our elephant's case it did not pay off. Shortly after the research was done, Lewis suffered multiple gunshots, very likely a result of crop raiding. He died in the Samburu reserve a year after the study was done."

The clash between elephants and people is as old as our species. To hold on in the long run, elephants need that precious commodity, land. Wildlife managers, says van Aarde, need to stop tinkering with elephant numbers by fencing herds in, culling in some places, and then supplying artificial water sources in others. Given enough room to move, elephant populations regulate themselves: they move on when they have fed too long in one area or when a water hole dries up. Elephants compete among themselves for the best habitats; the weakest ones are shunted off into marginal pockets of land, where they die off. This was the way of the Pleistocene world. Today few unfettered wildlife habitats anywhere on Earth are vast enough to hold what ecologists call a metapopulation of large mammals: a spread-out, shifting, breeding, dying group of animals, thriving in some places while it fades away in others.

Van Aarde contends that in parts of southern Africa, human population densities remain low enough that this ideal can become reality. There is a movement to create Transfrontier Conservation Areas in the region, protected wildlife habitats that stretch across political borders. In 2002, representatives of Mozambique, South Africa, and Zimbabwe signed an agreement to unite major conservation areas in the three countries, including Kruger National Park. Another proposed megapark, the Kavango-Zambezi, involves five countries and incorporates thirty-five national parks and reserves covering an area that harbors about a third of all of Africa's elephants.[20]

Local support is essential if this ambitious plan is to succeed. Van Aarde envisions some elephant hunting for subsistence or tourist income inside the megaparks but focused in areas that are not prime elephant habitat. Taking a controlled number of animals in population "sinks," where prospects for survival are low in any case, would not hurt the metapopulation.

Van Aarde and his colleagues have captured ninety-one elephants in seven south African countries and fitted them with GPS collars, and they are tracking their movements, looking for patterns that will help in the design of successful megaparks. So far, the data show that elephants prefer places where water is available and people are scarce, he says. Finding good forage is important, but not as critical as water and a bit of peace and quiet.

Why did elephants survive the Pleistocene die-off only in Asia and Africa? Paul Martin has argued that because the giants there had coevolved with people, they had time to adapt to human hunters. For many who study elephants past and present, however, that answer doesn't ring true. "That theory assumes that the animals are rather stupid and just kind of stand there while they're being killed," says Lister. "In reality, in places where elephants are in danger of poaching, they quickly learn to avoid humans, and they're very nervous."

The evidence for mammoths suggests that like living elephants, they were deeply social, long-lived, savvy, and able to communicate over long distances. Modern elephants speak to one another using infrasonic rumbles that carry far through the air, and they track the distant movements of their kind using specialized cells in their feet and trunks that can sense vibrations carried through the earth. Elephant herds have been known to flee when a group of their relatives were hunted down many miles away.

When people began to hunt America's ancient elephants, the great beasts must have understood quickly that the newcomers carrying the bladed sticks meant danger. But those were dry times. Mammoths and men alike were drawn to creeks and water holes, places where the grass stayed green. The need for water explains why, in the vast expanse of Pleistocene North America, there was no safe place for these giants to hide.

The continental United States holds more Pleistocene elephant kill sites than all of Africa, a land mass much larger than the United States, where people and elephants coexisted for far longer. All the American sites date to a narrow window of time, 13,000 years ago. Mammoths and mastodons were hunted from Arizona to Maine in that critical moment when Martin believes that humanity forever altered the face of the continent. Within a few hundred years, not only America's native elephants, but its horses, camels, dire wolves, saber-toothed cats, teratorns, and giant sloths would all be gone.

FIRST ENCOUNTERS

A man kneels on a bluff overlooking an oasis on Arizona's high, dry plains. Using a piece of deer antler to apply pressure, he drives flakes off a thin slice of chert braced against his thigh. Each movement of the antler tool is an act of skill and intense focus. The piece of pink-gray chert had been useful as a butchering knife until its edge broke against the sinews of the mammoth he and his comrades killed two days earlier. Now he is reworking the blade into a spearpoint. It is a difficult task, among the most delicate kinds of stonework his people know. The flakes fly off the chert, yielding a sharp new edge. But then one stroke falls wrong, and the well-worked piece of stone breaks in half. The flint-knapper stands and hurls the two useless pieces of chert away.

That moment of frustration is the easiest thing to imagine about this Pleistocene man, a maker of Clovis spearpoints, a hunter of bison and mammoth. We don't know what kind of clothes he wore or what his face looked like. We don't know if he was short or tall, young or old.

We do, however, know some strangely intimate details about this Clovis man and his clan. We know that he and his companions s killed a cow mammoth at the place now called Murray Springs, southeast of present-day Tucson. Most likely they ambushed her while she drank and ate at the oasis. During the long, grueling process of butchering her carcass, they set up camp on a hill overlooking the watering hole. We know that when the flint-knapper threw the broken knife, the two halves landed 23 feet apart. We know that in the process of killing the giant, someone in the group broke a spearpoint. The broken flake stayed with the mammoth's body, and the damaged point was left 262 feet away at the

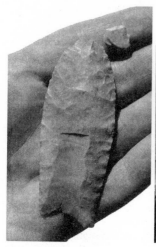

Figs. 13 and 14   A chip broke off this Clovis spearpoint when it was used to kill a mammoth at Murray Springs, Arizona (left). Mammoths drawn to oases in a time of drought were likely easy for hunters to find but still made daunting and dangerous prey. The mounted skeleton of a Columbian mammoth at the Page Museum in Los Angeles (right) gives a sense of what Pleistocene hunters faced. (Photo of spearpoint courtesy of Gary Haynes; mammoth skeleton photo from Wikimedia commons, en.wikipedia.org/wiki/ File:Columbian_mammoth.JPG.)

hunting camp. Today, it is possible to stand in Vance Haynes's lab at the University of Arizona and match the flake into its notch in the spearpoint, where it still fits with eerie precision.

Now in his eighties, Haynes, with his white hair and beard, thick reading glasses, and round belly, looks like Santa Claus. He sounds like the expert scientist he is, though, chatting easily about stratigraphy and the intricacies of the radiocarbon dating process.

Forty-two years after he first spotted mammoth bones eroding out of the arroyo wall at Murray Springs, Haynes's eyes still sparkle when he remembers the find or handles the many Clovis artifacts he and his colleagues excavated there. "Uncovering a Clovis point and realizing this has not seen the light of day for 13,000 years, that's something," he recalls. "It's what got me into archaeology."

The Murray Springs dig would reveal one of the most dramatic and best preserved of the few known Clovis kill sites. The frustrated flintknapper and his friends and relations had returned to the area on several occasions; they killed the cow mammoth there, and on other visits they surrounded and killed eleven bison (all calves and females) and may have taken down a horse.[1] Over the course of a few decades he and his people appear to have spent a lot of time in the San Pedro Valley, which holds, by Haynes's estimation, more than a fourth of all the known, stratified Clovis sites. (The remainder of the sites are scattered throughout the continental United States.)

The San Pedro has yielded an amazing array of mammoth and human remains. At the Lehner site, 10 miles south of Murray Springs, excavations begun in the 1950s revealed a group of twelve mammoths, along with thirteen Clovis spear points. Haynes helped uncover an ancient roasting pit there, where a rabbit and parts of a bear and a young mammoth had been cooked. A few miles east of Lehner, at Naco, Arizona, lay the bones of a single mammoth, which died pierced with eight Clovis spearpoints. The Naco mammoth may have been attacked along with the other Lehner animals but managed to escape, later bleeding to death. The remains of other mammoths have also been found nearby, though their association with Clovis people is not as clear-cut. Haynes has shown that all the San Pedro Valley mammoth sites date to approximately 13,000 years bp and posits that a single band of hunters may have been responsible for all the kills.

Today, a walk to the flint-knapper's campsite at Murray Springs passes through open desert, dotted with bushes of creosote and whitethorn. The seep that once sustained the oaks and the green grass that drew big prey animals in Clovis times is now a desiccated arroyo whose steep walls reveal layers of clay and sand. An obvious seam of dark earth runs midway along the wall—a layer of carbon-rich muck that buried each of the San Pedro Valley Clovis sites, sealing them against the ravages of time.

At Murray Springs, Haynes and his colleagues were able to peel off that layer of clay, which they dubbed the "black mat," as though it were a roll of carpeting. The surface beneath was so well preserved that mammoth tracks still pocked the ground around the cow mammoth's bones. Several feet off was a shallow well of the kind modern elephants are known to dig in Africa during times of drought. It seems clear that mammoths were a vital part of life for the people who killed and camped at the site so long ago—among the artifacts found was a heavy wrench carved from mammoth bone, most likely used to straighten warped spear shafts.

Before Haynes came along, most scientists had assumed that the first Americans had arrived 30,000 or more years ago. In the early 1960s, Haynes—a mechanical engineering student turned geologist and expert in radiocarbon dating—set out with archaeologist George Agogino to carbon-date as many verified Paleoindian sites as they could, in an effort that took them on a low-budget tour of the American West. Living on peanut butter sandwiches, they painstakingly gathered tiny grains of ancient charcoal for carbon dating. Sometimes it took weeks to get enough charcoal to make up a single sample: in those early days a tea-spoonful of material was needed to run a radiocarbon date. Soon Haynes was setting the clock for Paleoindian prehistory. In his lab, he pulverized part of a mammoth jaw from a kill site at Dent, Colorado, and measured the age of ancient plant parts that fell into the braincase of a butchered mammoth at Blackwater Draw, New Mexico, and of bits of charcoal that surrounded the mammoth that died of spear wounds at Naco. Every Clovis kill site he tested dated to between 13,500 and 13,000 years bp, much later than had previously been thought.

Around the same time, geologists Wallace Broeker and William Ferrin published work that showed a gap had opened in North America's glacial ice pack 14,000 years bp, clearing a passage from Alaska through Canada into the continental United States for the first time in many thousands of years. In 1964, Haynes published a paper in *Science* that pulled the evi-dence together into a tidy package. He suggested that the forebears of Clovis people had moved south through the ice-free corridor to north-ern Montana, bursting into the undiscovered landscape of the lower forty-eight states. It took them only a century or two to spread west as far as California and east as far as Maine.[2]

Haynes's vision was so persuasive that it quickly became a kind of gold standard in American archaeology. The new paradigm held that Clovis people, with their distinctive stone tools and their dramatic penchant for hunting mammoths, had been America's first settlers. Paul Martin was among the many researchers who latched on to this scenario, which con-veniently dovetailed with his theory of big-game hunters sweeping through North America and wiping out Ice Age megafauna.

Archaeologists nevertheless continued to come up with finds that they believed proved a much earlier human presence; in many cases Haynes found evidence to disprove these claims. Purported artifacts at Calico

Hills in California, which luminaries including Louis Leakey believed showed evidence of toolmaking by early humans hundreds of thousands of years ago, turned out to be bits of rock broken apart by erosion. In other cases, such as the Meadowcroft Rock Shelter in Pennsylvania, the sites showed clear remnants of human occupation but the radiocarbon dates were dubious.

After decades of scrutiny, however, there remain a few sites that, depending which expert you ask, could be considered as having broken the "Clovis barrier." At Cactus Hill in Virginia and the Topper site in South Carolina, stone tools have been found stratified below a layer of Clovis artifacts, in sediments that date to pre-Clovis times. The bones of apparently butchered mastodons at the Schaefer and Hebior sites in Wisconsin have been radiocarbon dated to more than 14,000 years bp.

Many scientists now acknowledge that Clovis people were not the first to live in the Americas. The most convincing evidence of this comes from sites thousands of miles apart, one in Oregon's arid interior and another near the coast of southern Chile. The critical artifacts are not big, elegant spearpoints but wooden tools, the remnants of a 12,000-year-old mastodon-skin tent, ancient bits of seaweed stuck to a crude bone knife, and a scattering of miraculously well-preserved Pleistocene human feces.[3]

At Monte Verde in southern Chile, a team led by Tom Dillehay of Vanderbilt University excavated the remains of a human camp. Some of the edible plants they found were not endemic to the area and must have been carried to the site from miles away. They also uncovered an assortment of mastodon and other animal bones, including a mastodon tusk that looks to have been reshaped by human hands. Wood from the site radiocarbon dated to 14,500 bp, a thousand years earlier than the Clovis dates.[4] The seaweeds, well-preserved examples of species that are eaten or used as medicine by local indigenous people in modern Chile, date to the same window of time.[5]

Monte Verde lies at the tip of South America, about as far from the Bering land bridge as it's possible to get. If people were living near Santiago a thousand years before Clovis, they could not have traveled through the ice-free corridor that Haynes claimed formed the gateway from Beringia to the Americas, because that pathway would have been under a vast river of ice. Not only was the Monte Verde site surprisingly old, but it lacked elements that experts on early American settlement expected to find. The few stone tools found were simple modified pebbles, nothing like the elegant bifaces made by Clovis people. Although Haynes remains skeptical, Monte Verde triggered a dramatic change in many researchers' view of American prehistory, shattering the image of Clovis first (and indirectly calling into question Paul Martin's overkill model).

Clovis spearpoints are easy to identify: big, leaf-shaped bifaces of obsidian or chert, with characteristic grooves, called flutes, running up the center of the base. The flutes made it easier to bind the point to a spear shaft and are a uniquely American invention. People who came out of northeast Asia and made an eastward crossing of Beringia would eventually give rise to the roving bands of foragers that hunted at Murray Springs and other Clovis sites. But the stonework of Beringia was markedly different from that of Clovis, and the first people to enter the New World made their living very differently than Clovis people later would.

The traditional image of America's first pioneers was that they were big-game hunters who honed their skills on woolly mammoths and other Ice Age giants as they made their way across Beringia. Martin promoted this view, describing "skillful, robust, accomplished, highly adaptable, and above all, persistent and very likely passionate hunters.... In the process of crossing the Bering Land Bridge to the New World, the early Americans must have found many animals familiar to them, such as woolly mammoths [and] musk oxen."[6] Martin never made a serious study of Beringian archaeology, however, and the researchers who have paint a very different picture of the Clovis people's ancestors and their hunting habits.

Humans penetrated western Beringia long before they reached America, living above the Arctic Circle at the mouth of the Yana River in far northeast Asia 27,000 years bp. They left behind some spectacular artifacts, including a spear foreshaft (a graceful clamp that joined the spear to the sharp projectile point) made from the horn of a woolly rhinoceros, and a scattering of stone tools.[7] When the cold days of the last glacial maximum came, people abandoned the far north. Beringia would not be resettled until 15,000–14,000 years bp, when the glaciers were in retreat.

The earliest archaeological sites in central Alaska—dating to pre-Clovis times—were found at Swan Point in the Tanana River Valley in the early 1990s. The tool kit from the lowest level at Swan Point is distinctly not Clovis but looks identical to the artifacts of the late Pleistocene people of the Lena Basin of Siberia, across the Bering Strait.[8] The settlers at Swan Point made spearpoints and knives out of antler or ivory, slotted the blades along each edge, and studded them with tiny, razor-sharp flakes of obsidian. Archaeologists call this microblade technology. If your idea of Stone Age tools is based on common Paleoindian arrowheads, a microblade spearpoint looks, by comparison, like the work of some prehistoric obsessive-compulsive. In Siberia, slotted points have been found embedded in the bones of bison, the sharp microblades still studding their edges.

Why would hunters go to all the effort of making a slotted point instead of simply chipping an entire spearhead out of stone? Solid stone tools could not stand up to the frigid conditions of the far north: they would shatter on impact. Yet flexible materials like bone, antler, or ivory do not hold a sharp edge. Microblades were a solution to this problem, creating a lethally sharp yet flexible point.[9] People in the north carried wedges of chert or obsidian from which tiny microblades could easily be flaked, and they fashioned their multimedia spearpoints wherever they traveled.

Like their contemporaries in Siberia, the early settlers of the Tanana Valley hunted elk, reindeer, snowshoe hare, and geese, which along with other waterfowl became increasingly plentiful as the landscape grew wetter. They also fished the rivers. Mammoth hunting, though, was not part of their culture, and they likely never saw a living specimen. Woolly mammoth ivory can be found in their tools, since it remained a plentiful resource long after the animals died out. Even today mammoth bones litter the Beringian landscape, and every year new fossil finds, propelled by the movement of melting and refreezing ice, come heaving out of the permafrost.

The microblade-point makers of the northwest coast were able seamen, and their diet included an abundance of seafood. At On Your Knees Cave, on Prince of Wales Island in southeast Alaska, two of these ancient people were found buried, their tools laid in the grave with them. Their bones date to 12,500 years bp, and analysis of the nitrogen isotopes in the remains show that they were successful fishers and hunters of marine mammals.[10] A diet of seafood shifts the balance of nitrogen isotopes in a body, and the pair found at On Your Knees Cave had the same isotope signature as that of a ringed seal or sea otter.

Glacial ice had receded from much of the Pacific Northwest coast by 15,000 years ago. At that time, the ice-free corridor leading from the Alaskan interior through Canada and into the lower forty-eight states had not yet opened and would not do so until 13,500 years bp, just before Clovis time.[11] James Dixon, an archaeologist at University of Colorado–Boulder, argues that the first Americans migrated not across the interior of the Bering land bridge but along its southern coast. Black and brown bears, which, unlike humans, left plenty of fossil clues behind, moved into North America along this route. If these big omnivores could thrive along the Beringian coast at that moment in time, humans would have been able to make it, too.[12] People would have thrived on shellfish and seal meat, their knowledge of coastal survival strategies serving them well as they moved south along the American shoreline. Since they knew how to harvest the bounty of the sea, they likely paid large land-based targets like mammoths and ground sloths little attention. The earliest evidence of watercraft in North America is a pair of human thigh bones that date to 12,900 years bp, found at Arlington Springs on Santa Rosa Island in what is now the Santa Barbara Channel off the Southern California coast.[13] The Channel Islands were never connected to the mainland during the late Pleistocene; the only way to get to Arlington Springs would have been by boat.

If the first Americans migrated down the western coast, that explains how they reached Monte Verde in southern Chile a thousand years before the dawn of Clovis culture. Not only was the way open earlier, but the coast formed a sort of biotic highway that would allow people to adapt rapidly as they moved south.

Rising seas have buried most evidence of early coastal people in the Americas. And no trace of Pleistocene people has ever been found along the theoretical ice-free corridor, either. Yet very old human bones have been uncovered at the Anzick site in Montana, at the southern end of the corridor, as they have been at Arlington Springs and On Your Knees Cave. Both models could be correct: it is possible that the earliest American pioneers traveled south along the coast, and that centuries later a new wave of colonists, makers of solid-stone tools similar to Clovis points, began to come south and east out of the Alaskan interior, moving toward the Great Plains.

Recent studies of DNA from ancient Paleoindian bones, and from living native people, offer new clues to the movements of America's first settlers. One unique genetic marker, D9S1120, is widespread among ancient and modern Native Americans and in a few groups of indigenous Siberian people but has been found nowhere else on the planet. The simplest explanation for this pattern is that a single founding population moved out of Siberia and passed the gene down to all the diverse groups of Native Americans. Whoever populated the Americas first, in other words, populated it well.[14]

This genetic evidence resonates with an idea that some archaeologists have put forward: that the rapid spread of Clovis spearpoints throughout the continental United States 13,000 years bp reflects the adoption of a new technology among human populations already scattered across North America, rather than the sudden expansion of a single group of people. The Clovis point was such a hot new trend that it jumped from clan to clan like a prehistoric Hula-hoop.

The earliest human populations in America were likely few, and scattered widely over the landscape. Still, necessity would have brought separate groups together on occasion, to trade and to find mates. The invention of the Clovis point could have spread rapidly through such established trade routes.[15] Millennia after the Clovis culture faded, another new invention, the bow and arrow, would spread across all of North America in the archaeological blink of an eye. No one argues that a single group of bow-and-arrow hunters conquered the continent; it is clear that the new weapon spread quickly through existing trade and social networks.

Martin remained stubbornly attached to the Clovis-first model and saw it as critical to the theory of human overkill in America. Yet from the mammoth's point of view, it matters little whether Clovis people were the first to settle in the New World, or if they were members of a single cohesive culture. Under siege in a time of rapidly shifting climate, America's native elephants were hunted, coast to coast, by people who wielded spears tipped with big, deadly, razor-sharp stone points. These hunters succeeded in bringing down the great beasts, at a moment when they appear to have been unusually vulnerable.

The Murray Springs mammoth kill stands as a classic example of that scenario. More than four decades after the excavation there was completed, Vance Haynes continues to uncover its lessons. In May 2008, he published a new study summarizing evidence that the last of the mammoths died as Clovis culture spread through North America.[16] At the same moment, the continent's last wild horses, camels, tapirs, and dire wolves disappeared. The planet was warming then, and on Arizona's high plains, the water table dropped, forcing mammoths to dig for water. The pattern is evident at Murray Springs, Lehner Ranch, Blackwater Draw, and other Clovis sites in the arid Southwest. But Haynes has found similar evidence at the Hiscock mastodon site in New York and the Domebo mammoth site in Oklahoma as well.

Across the continent, Clovis people left their spearpoints in big animals that they encountered near water sources during a few hundred warm, dry years. Then the climate flipped back to frigid, glacial conditions, in a final Pleistocene cold snap known to paleontologists as the Younger Dryas period, named after the tundra wildflower *Dryas octopetala*, whose pollen is commonly preserved in soils of the time. Low temperatures reduced evaporation and raised water tables, the waters ran again at Murray Springs, and lake and pond levels rose everywhere. That wet, chill time is recorded in the black mats, formed from wetland plants that rotted to dark clay and covered, then preserved, the Clovis kill sites. But for the Pleistocene giants, the abundant water came too late. No remains of the vanished megafauna have ever been found in the black mat clays, or in layers above them. Neither has a single Clovis point.

People survived, but the points they made became smaller and lighter. The spearpoints made by Folsom people are differently shaped and more compact than Clovis points but share the distinctive fluting at the base. At several archaeological sites in the western United States, Folsom artifacts, along with the bones of butchered bison, are found in black mats overlying deposits of mammoth remains and older Clovis tools. Folsom society likely grew out of Clovis culture, an adaptation to a landscape emptied of all its largest beasts except for bison.

"If you find a mammoth tooth up against the black layer, that's the last mammoth on that part of the planet," says Haynes. "The black mats in Arizona, Nebraska, the Dakotas, all date to 13,000 years ago. So do the human bones from Anzick and Arlington Springs, the oldest human remains on the continent." He has spent much of his life exploring that brief, long-ago moment when Clovis culture blossomed and the last mammoths vanished. It is a compelling story that seems to beg for a single dramatic explanation, but Haynes claims he is not attached to any particular theory.[17] "Something very interesting happened here 13,000 years ago," he says, grinning over the rims of his reading glasses. "We don't know yet what it was, but I'd say, stay tuned."

America is not the only land where the arrival of Pleistocene people appears linked with a devastating wave of extinctions. Half a world away from Murray Springs, researchers ponder fossils and sediments gathered from the remote Australian outback, trying to puzzle out the lives, and mass disappearance, of beasts much stranger than America's lost giants. Looming marsupials ruled Australia until the first people arrived there, about 50,000 years bp. Those ancient settlers did not carry or make stone spearpoints. Yet they still managed to transform the continent, using only wooden tools and their bare hands.

GIANTS
DOWN
UNDER

PREHISTORIC PITFALLS

The great hunter waited in ambush, high enough in the eucalyptus that she could watch her target unnoticed. She bided her time while the object of her attention—a browsing kangaroo—moved slowly closer, feeding on the leaves of nearby trees. At the last possible moment, the hunter pounced. Their bodies rolled on the ground, hearts beating fast: the kangaroo, the marsupial lion, and the cub she carried in her pouch. Then abruptly, all three fell into emptiness. After a long, panicked drop through the abyss, they slammed into the ground. Perhaps they died on impact. If they survived for a time, the Ice Age creatures would have found themselves surrounded by bones, in a cave lit dimly by the glow from the small opening far above.

A four-hour drive east of Adelaide, outside the small South Australia farming town of Naracoorte, is a subterranean complex of limestone caverns. Today, visitors can follow a walkway through an underground tunnel and emerge in the great chamber where the predator and her prey fell together to their deaths, tens of thousands of years ago. A mounted skeleton of *Thylacoleo*, the extinct marsupial lion, stands in the foreground. Behind it, a floodlit jumble of unexcavated skulls, jaws, ribs, and limb bones stretches into infinity.

Julian Edmund Tenison Woods, a Catholic priest and devoted naturalist, first described the Naracoorte caves in 1859. Woods had at first believed that the sort of bone deposits he found at Naracoorte were evidence of a biblical deluge. In light of Charles Darwin's controversial new theory of evolution, he saw them for what they were: a rare record of past life forms, many of which had vanished forever.[1] He recognized that he was looking at the remains of extinct animals, closely related to some creatures that still roamed in the region, but much larger than their living counterparts. All had stumbled to their deaths through small openings that led into the great underground caverns.

Blanche Cave, which Woods studied, became famous for its high ceilings and fantastic limestone spires, but its fossil deposits were, for the most part, ignored. It became a popular site for balls and other Victorian celebrations, and the deeper scientific questions of the caves stood neglected. The caves lay unexamined until the 1950s, when enthusiasts of the new sport of spelunking began to explore them, using Woods's book *Geological Observations* as a guide.

One of those cavers was a blue-eyed young man named Rod Wells. He was working as an engineer, spelunking in his free time, when he joined a University of Sydney caving expedition in 1963. Inspired by the professional mammalogists and paleontologists on the trip—and his childhood rambles in the Blue Mountains, spent studying insects, mammals, and rocks—Wells quit his engineering job, moved to South Australia, and began a program in zoology at the University of Adelaide.

He began to spend weekends searching for marsupial fossils in the caves that had eroded millions of years earlier on the limestone coast. At the time, only a few examples of extinct marsupial giants had been identified, and little was known about them. Often the identification of an entire vanished species rested on one or two skulls or jawbones.

After hearing about the odd-shaped, sharp-clawed hand of a marsupial lion that had been reassembled from scattered bones found in a quarry at Naracoorte, Wells turned his attention to the little-explored caves there. He put in long days, crawling through narrow openings, rappelling off subterranean ledges by the dim light of a headlamp. On October 29, 1969, Wells was with his caving buddy, Grant Gartrell, when Gartrell felt a breeze wafting from a rock pile. They began prying rocks out of a passage so narrow that the men had to remove their helmets to squirm through it.

They found themselves in a large cavern, floored with silt and what looked to Wells, in the weak light, like odd, jagged rocks. It slowly dawned on him that they were looking at the upturned skulls of extinct kangaroos. These would be confirmed as the first complete specimens ever found. In the chamber later dubbed Victoria Fossil Cave, the first-known complete skeleton of a marsupial lion was also lying at his feet.

More than forty years after its discovery, Victoria Fossil Cave remains the richest deposit of Pleistocene fossils ever found in Australia. Wells and his colleagues have so far managed to sort through about 4 percent of the 5,000 tons of fill in the chamber, from which they have identified more than 5,200 individual animals representing 108 species. Of the creatures whose bones have been catalogued, about half still exist in parts of Australia today; the other half, including some of the strangest mammals that ever walked the earth, have vanished forever.

An open hole, filled in long ago by shifting dirt and rock, once led from the surface into the underground cavern, acting as a pitfall trap for Australia's Pleistocene mammals. Isolated from other land masses for millions of years, the continent grew its own distinctive array of marsupial giants: *Diprotodon*, a rhino-sized herbivore with clawed toes and beaverish buck teeth; *Procoptodon*, a towering, pug-faced kangaroo that reached high into treetops to feed; *Macropus titan*, a scaled-up version of the living gray kangaroo; *Thylacoleo carnifex*, the marsupial lion, a predator as big as its African counterpart but with opposable thumbs, each tipped with a claw that was sharp as a switchblade. There was *Megalania*, a monster lizard, and *Wonambi*, a 20-foot-long snake that waited in ambush near water holes and then suffocated its prey in the coils of its massive body.

The discovery of Victoria Fossil Cave opened a new chapter in Australian paleontology, drawing international attention. Even Paul Martin, who was then promoting his theory of human overkill, made a pilgrimage to Naracoorte to examine the find. Martin had suggested that the American blitzkrieg happened so fast that little fossil evidence had been left behind, and took the scarcity of butchered megafauna remains as confirmation of his model. Wells, who had put in so much time searching through unexplored tunnels deep underground, was unimpressed with that argument. "You might as well sit on the beach in Hawaii," he wisecracked to Martin. "You won't find evidence there, and that will support your hypothesis."

The paleontology lab that Wells shares with other researchers at Flinders University in Adelaide is a fossil fanatic's dream, full of cardboard cartons heaped with long, yellowed bones, an assortment of skulls (from the massive to the dainty), and numerous mandibles and loose teeth. Hunched over the lab benches, three graduate students painstakingly match stray humeri with the correct lower arm bones and fit incisors into the deep sockets where they formed long ago. Gavin Prideaux—Wells's former student, now a prominent paleontologist in his own right—had just returned from a trip to Tight Entrance Cave, a remote site in southwestern Australia, and brought these new finds back with him. Several *Diprotodon* skulls lay among the bones; the creature has proven to be ubiquitous at fossil sites scattered across the continent. Holding a molar in his palm, Wells could tell at a glance that a *Diprotodon* had grown the tooth. The only question was which of the many excavated skulls it belonged to.

Diprotodon was first described in 1838 by Richard Owen, a British biologist and founding father of paleontology, who would also classify giant sloths and other South American megafauna collected by Darwin. Owen's first glimpse of *Diprotodon* was of a fragment of lower jaw that

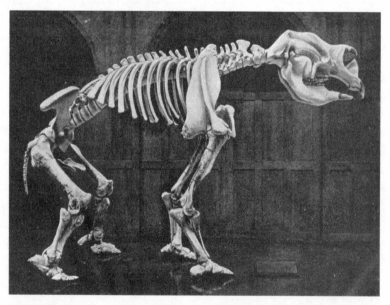

Fig. 15 *Diprotodon*, a rhino-sized, bucktoothed vegetarian, was the largest marsupial ever to walk the earth. The beast was widespread in Pleistocene Australia but disappeared soon after the first human settlers arrived. (Photo courtesy of www.search4dinosaurs.com.)

contained the stub of an enormous incisor, found at Wellington in New South Wales. Though he was deeply impressed by the magnitude of these bones, it turned out that this first fossil was from a juvenile. When Owen finally got his hands on adult *Diprotodon* specimens (a femur and some molars), he at first mistook them for elephant bones.[2]

Paleontologists today still struggle to understand what kind of creature *Diprotodon* was and how it lived. The biggest marsupial ever to have walked the earth, an adult stood 5.9 feet high at the shoulder, stretched nearly 10 feet from nose to tail, and weighed more than 5,500 pounds. Different researchers over time have classified their *Diprotodon* finds as several separate species, some distinctly larger than others at adulthood. But in a 2008 paper, paleontologist Gilbert Price argued that there had been only one species of *Diprotodon*, in which males grew much larger than females.[3] Such differences in stature between sexes are common among big, herbivorous, placental mammals—think of elk, bison, or domestic cattle—and also in some of the largest living marsupials, such as red and gray kangaroos.

In Price's vision of *Diprotodon*, the big beasts lived in herds. A male would have mated with several females and may have protected his harem from rivals, as wild horses do today. However its social life was structured, *Diprotodon* thrived in arid conditions and fed on all sorts of vegetation. Its huge incisor teeth, worn down by continual gnawing on tough leaves and twigs, grew throughout its life, as a beaver's do.

The quantity of specimens at Victoria Fossil Cave enabled Wells to uncover a wealth of information about some lesser known Pleistocene giants. Among these was the marsupial lion, *Thylacoleo*. Its Latin name means "pouched lion," but many researchers questioned whether it had been a predator at all. For one thing, the beast was clearly descended from the same long line of herbivores as *Diprotodon* and the wombat, a living marsupial that digs underground burrows, eats grass, and looks like a cross between an overgrown rabbit and a teddy bear. How could Australia's greatest mammalian carnivore have descended from a lineage of cuddly vegetarians? To complicate matters, *Thylacoleo* possessed perhaps the oddest set of teeth ever evolved. Paleontologists pay close attention to fossil teeth—they are often the best preserved remains of an animal, and many species can be identified solely from the unique number and shape of the teeth. For creatures like *Diprotodon* and

Thylacoleo, with no living analogs, teeth provide critical clues to the way the vanished animals once lived.

A good way to begin to grasp how much information is encoded in these ancient choppers is to look at, or feel, your own set. Humans have the same classes of teeth as other mammals: incisors (the shovel-shaped front teeth), canines (nondescript in people, a sharp, stabbing weapon in the mouth of a wolf or lion), and premolars and molars (the broader, cusped teeth that line the inside of our cheeks). Animals evolve specialized sets of teeth for their particular diet. Cows, bison, and other ruminants have no upper incisors—they use their lower front teeth to tear grass, and a long array of wide, tough cheek teeth to grind it up. Of all the foods on earth, grass is toughest on teeth. Each leaf is filled with abrasive grains of silica, so a grazer's teeth must withstand chewing on the biological equivalent of sandpaper. Predators, on the other hand, tend to have strong, prominent canine teeth, the saber-toothed cat being an extreme example. In wolves and other members of the dog family, some of the molars and premolars have turned into jagged knives, called carnassials, designed to shred meat. They have kept some cheek teeth for grinding, useful for gnawing on bones or grazing on plant foods. Cats, large or small, are the ultimate carnivores: they have lost most of their cheek teeth, except for two or three carnassials that slice against each other, ripping meat away from tendon and bone. A cat's mouth was made to eat meat and little else.

In Victoria Fossil Cave, where a spotlight shines on the mounted *Thylacoleo* skeleton, one can examine the marsupial version of tiger teeth. This beast had dispensed with the series of powerful molars its plant-eating ancestors grew. The equipment remaining is simple, but startling to anyone who has studied mammalian dentistry. At the front of the mouth stand two pairs of fearsome fangs, big pointed incisors perfect for stabbing and grabbing. Most of the cheek teeth have vanished, but a single premolar remains, having morphed into a long, serrated knife blade. *Thylacoleo*'s slashing premolar looks man-made, as if a Stone Age artisan had chipped a knife of chert and set it in the animal's mouth.

Earlier researchers assumed—based on limited evidence—that *Thylacoleo* was a fruit eater, as its teeth were clearly not designed to handle tough shrubbery or grass. The vast bone deposits at Naracoorte made it possible, for the first time, to study skulls and jawbones of

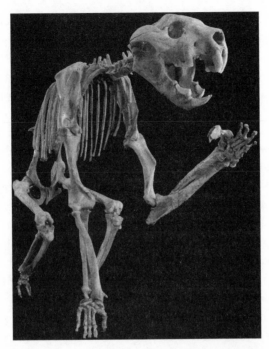

Fig. 16   A skeleton of a marsupial lion, *Thylacoleo carnifex*, Australia's largest-ever mammalian predator. *Thylacoleo* preyed on giant kangaroos and other long-extinct megafauna. (Photo by Rod Wells.)

numerous *Thylacoleo* specimens, young and old, and to examine the ways their teeth had changed over their lifetimes. Wells and his colleagues learned that infants grew a formidable set even before they had left the mother's pouch and that adult teeth wore down in predictable patterns—though they kept some sharp edges. There was plenty of evidence that adults ate meat, including microscopic wear patterns on adult teeth that resembled those found on known carnivores. To drive the point home, Wells put his engineering skills to work and built a machine that imitated the motions of a *Thylacoleo* jaw, fitted with imitation premolars made of steel. First he fed his mechanized monster eucalyptus leaves, then kangaroo meat and hide, and afterward compared the marks left on the artificial teeth by each kind of food. Meat left wear marks that closely resembled those on the actual fossil teeth.[4]

According to Wells, the creature's sharply clawed thumbs were "semi-opposable," almost, but not quite, as flexible as the human version. The thumb would have helped *Thylacoleo* subdue struggling prey and also grasp branches as it climbed.[5] The beast preyed on the strange, extinct forms of kangaroo whose bones line the floor of Victoria Fossil Cave, and like a leopard, it hauled its victims up trees to eat in peace. There it could avoid interference—and theft—from marsupial tigers, giant Tasmanian devils, and outsized lizards, the other large carnivores of the era.

The most common megafauna bones in the Naracoorte caves are those of short-faced, or sthenurine, kangaroos, a family of creatures that flourished during the Pleistocene but had disappeared by its end. The biggest kangaroo ever to exist, *Procoptodon goliah*, was a sthenurine that stood just under 10 feet tall. Its mounted skeleton looks like a fanciful invention, as if someone had stacked the upper body of an early hominid atop the hips and legs of a kangaroo. Instead of the foreshortened arms of a modern kangaroo, *Procoptodon* had eerily human upper limbs. Its face was dramatically different from any living kangaroo's, too: massive and square, with a short snout like a bulldog. The skull was built for powerful grinding, needed to break up leaves and branches. The eyes faced forward, like a human's, giving the animal stereoscopic vision, a trait usually found in predators, who need good depth perception to track prey. *P. goliah* was probably adapted to reach up into treetops to browse on leaves: its skull shape and teeth closely resemble that of the koala, which feeds exclusively on eucalyptus leaves. Forward-facing eyes and stereoscopic vision would have helped it find and reach the choicest foliage.

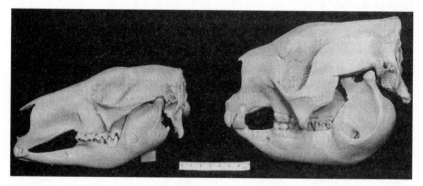

Fig. 17 Skulls of modern kangaroos (left) and extinct short-faced kangaroos (right) reflect their different lifeways. Living kangaroos are grazers with eyes set on the sides of the face, providing wraparound vision that helps the animals keep watch for predators. The short-faced kangaroos had powerful jaws suited to feed on tree leaves and twigs. Both eyes faced forward, giving the creature good depth perception to aid its ability to find and select the freshest foliage. (Photo by Rod Wells.)

Surviving kangaroo species all have deerlike faces, with a long, slender snout and eye sockets on each side of the head, providing the wraparound vision that helps most grazers keep watch for predators. With adult males weighing 140 pounds or more, the red kangaroo, Australia's largest surviving native mammal, is petite compared to *Procoptodon*. And, like all the living forms of kangaroo, it feeds primarily on grass.

Wells discovered an important pattern in ancient teeth that shed light on Australia's mass extinctions. Most of the vanished herbivores were browsers, which feed on shrubs and trees, rather than grazers, which rely on grass. Their jaws were powerful, built to deliver the force required to chew up branches, but only *Diprotodon* had teeth that could withstand the nonstop abrasion of eating grass. Most true grazers have some special adaptation to compensate for the extreme toll their diet takes on their teeth. Modern kangaroos, for instance, replace their molars as they wear down. Throughout its life span, new cheek teeth grow in and move forward along the jaw, as worn ones drop off. The disappearance of the great browsers suggests that a shift in dominant vegetation, from shrubs and trees to grasses, was involved in their demise.

Australia's Pleistocene giants evolved on a drying continent. For eons, Australia was linked to Antarctica and South America in a great land mass known to geologists as Gondwanaland. (In fact, Australia's first marsupials originated in South America and crossed Antarctica to reach their new home.) The island continent broke loose about 45 million years ago and began drifting north into warmer latitudes. During the Pleistocene, glacial cycles in the northern hemisphere drove climate changes far to the south. When glaciers peaked, global temperatures dropped, slowing evaporation of water from the oceans to the atmosphere and choking off the monsoonal rains that in warmer times watered the Australian interior. About one million years ago, Australia experienced a lasting dry spell that altered all its habitats. Great forests that had once covered much of the continent collapsed, giving way to grassland and open woodland. *Diprotodon*, *Procoptodon*, *Thylacoleo*, and many of their contemporaries were arid-adapted creatures; their large size may have helped them survive in times of severe drought.

Yet increasing aridity may ultimately have driven all the great beasts into oblivion, according to Wells. Australia's massive quadrupeds, like

Diprotodon and *Thylacoleo*, were not built to move far or fast. Anchored to shrinking water sources in times of drought, they would have died slow deaths. This is a pattern that has been documented repeatedly, even among modern, highly mobile kangaroos. Drought killed thousands of kangaroos in South Australia's Flinders Ranges in 1969 and again in the early 1980s. As the landscape dried, most mammals, which can only travel as far as a stomach full of water will take them, became stranded at a few remaining water sources. Unable to move far enough to find food, they gathered at water holes, and ultimately died there. Wells recalls that at one small water hole, "about the size of my office," he counted a total of 80 red kangaroo carcasses.

While the Australian megafauna had survived previous dry spells as northern glaciers waxed and waned during the Pleistocene, with each glacial cycle conditions on the continent appear to have became harsher. The oldest fossil deposits at Naracoorte show that more than 400,000 years ago, the area enjoyed relative abundance, with plenty of forage for herbivores. Wells thinks that during each subsequent glacial maximum plant life became less plentiful, as the climate became colder and drier. "The system never rebounds to its full productivity again before another glacial comes along," he explains. Wells's scenario is unproven, however—he still needs evidence to back it up. His hypothesis is based, in part, on data from formations, called speleothems, at the Naracoorte complex and other Australian caves. "Speleothem" is a catch-all phrase that describes stalactites, stalagmites, and flowstones that form in limestone caves. These are created when water dissolves minute amounts of the calcium carbonate that makes up the limestone walls and deposits it as calcite within the cave. The deposits take an array of fantastic forms: cones, straws, curtains, or pillars that grow up from a cave floor, down from its ceiling, or out from its walls. Wells helped pioneer the use of speleothems as a way of tracking ancient climates around Naracoorte and correlating them with fossil deposits of different ages.

Speleothem deposition is proportional to rainfall outside a cave: not the total amount of water that falls from the sky, but the amount that soaks into the soil. In arid conditions, much of the rain evaporates before it can percolate below-ground, and speleothems do not grow. Cave formations thus provide an index to the changing climate around Naracoorte throughout the late Pleistocene. In Cathedral Cave at Naracoorte, Gavin

Prideaux has tracked matching sequences of bone deposition and speleothem development.[6] The shifting populations of local animals reflect the drop-off in effective rainfall shown by speleothems, he found. In arid times, populations of animals adapted to forest and closed woodland wane. Others better adapted to dry conditions, such as *Procoptodon goliah*, become more abundant.

Wells and Prideaux agree on the overall pattern of the Naracoorte evidence: it shows that Australia's climate shifted in cycles related to the state of northern hemisphere glaciers and that the habitats outside the cave, and the megafauna that depended on them, responded to those changes. Yet they draw opposing conclusions as to the cause of the megafauna's demise. Prideaux believes that human impacts drove the extinctions, whereas Wells considers climate to be the deciding factor (though he is willing to allow that humans might have helped tip the balance). Still, Wells points out that there is no solid evidence of ancient people ever hunting megafauna in Australia.

The first humans reached Australia by 45,000–50,000 years bp.[7] The question of whether, and how, Australia's first people affected the megafauna there has been debated for decades. As in the Americas, one camp argues that the giants were done in by climate change, while another sees humans as the cause. In the 1960s and 1970s, the problem was intractable because the youngest known megafauna bones were more than 30,000 years old, exceeding the age limit for reliable use of the radiocarbon technique. Recent refinements in the treatment of samples have made radiocarbon dating usable on older materials, but the more ancient the find, the more doubtful the method's accuracy. Luminescence dating, which measures the time elapsed since artifacts or sediments were last exposed to sunlight, can date much older sediments but has its own technical problems.

In 2001, a team of researchers led by Richard Roberts of the University of Melbourne published a paper in *Science* reporting a striking pattern that emerged when they dated late Pleistocene megafauna bones from twenty-eight sites across the continent.[8] Roberts and his coworkers claimed to have found solid evidence that Australia's megafauna had died out quickly, in a variety of different habitat types and climate zones, about 46,000 years ago. The two youngest samples, which dated to 47,000 and 46,000 years bp, came from Queensland, on Australia's eastern edge, and from Western Australia, on the opposite side of the country. The widespread die-off came well before the last glacial maximum of the Pleistocene hit its peak, at a time of relatively moderate climate. The findings corroborated the idea that newly arrived humans drove the extinctions.

Many found the Roberts paper convincing, and it has been widely cited by paleoecologists, but the debate surrounding the megafauna extinction Down Under is far from over. Prideaux's approach is to look far back into Pleistocene time, well before Australia's giants died out, to try to understand how they handled climate stress during earlier glacial cycles. The Naracoorte caves make an excellent starting point—because the cave chambers opened to the surface and then closed again at different moments during the Pleistocene, they captured snapshots of local wildlife populations under different climatic conditions.

The Cathedral Cave deposits Prideaux studied at Naracoorte span a relatively wet time about 300,000 years bp, a shift to arid conditions during a glacial maximum 270,000 years bp, and the later onset of a new wet phase. Populations of numerous animals waned during the glacial maximum but later bounced back. Biodiversity seems to have dropped during arid times, but some megafauna thrived: *Procoptodon gilli*, for example, a short-faced kangaroo that preferred dry heathland, increased in numbers. The megafauna cannot be treated as a single entity: different species thrived in different kinds of habitat. In Prideaux's view, no existing hypothesis explains how climate change could have wiped out these diverse megafauna all across Australia, at the same moment in geologic time.

Prideaux has also worked on a spectacular fossil find in the Nullarbor Plain, one of the harshest deserts of the Australian interior. The evidence from its caves includes beautifully preserved bones of marsupial lions and a variety of kangaroos, including several extinct species that had never before been identified. Isotope analysis of the fossils, which reflects the animals' diets, shows that these 400,000-year-old beasts were adapted to life in an arid climate.[9] This, Prideaux argues, undermines the argument that the megafauna were unable to handle increasingly dry conditions. In fact, the majority of the extinct species found in the Nullarbor caves had been adapted to arid habitats for half a million years.

They survived late into the Pleistocene, thriving in a vast desert, and always bounced back after times of drought—until humans arrived.

Wells remains unconvinced by Prideaux's theory. As intriguing as the climate record from the Naracoorte caves is, he argues that it only shows general trends, missing details that could have spelled life or death for many animals. "If we're dealing with biological extinction, we need to start thinking in terms of biological time frames, rather than geological ones," he says. To illustrate the subtle, short-term changes that can have disastrous effects on wildlife populations, he offers an example from early in his career, when he was studying the life history of the southern hairy-nosed wombat.

Wombats mate in autumn, and the young begin to leave their mother's pouches in spring or early summer. In a good year, there is enough early rain to water a spurt of new grass in time for the youngsters to find nutritious forage. When that happens, young wombats thrive. If the rains are late or minimal, however, most of the year's new brood of wombats starve. Those that survive won't start to reproduce until they reach three years of age. When Wells studied rainfall records covering the previous twelve decades, he found that the period had included only twenty years with sufficient rain to bring on a strong growth of grass in the critical season. Most of the time, the wombat population was static, struggling to hang on. Given these conditions, says Wells, it makes sense that adult animals live twenty-five years or more: they may successfully reproduce only once or twice during that span. A minor change in the timing of rainfall is enough to send a population plummeting. Giants like *Diprotodon* and the short-faced kangaroos would have reproduced much more slowly and been even more vulnerable to aberrations in the climate.

Dating samples in Australia has proven tricky, so it has been difficult to figure out when the giants disappeared. No one really knows when the last of the megafauna died out, though as more sites are dated, all the youngest date to 40,000–50,000 years bp. Because the organic matter needed for carbon dating quickly leaches out of bone in Australia's acid soils and hot, dry climate, most researchers measure the age of their fossil finds by applying luminescence or radiocarbon techniques to surrounding sediments. All too often, though, sediments shift over time. Bones are washed away from the sites where animals actually died; older remains are jumbled up with younger ones. One of the few sites on the continent where megafauna bones have been found together with human artifacts is at Cuddie Springs in New South Wales. The site has been dated to about 30,000 years bp, but the significance of the evidence is still unclear, as many scientists have dismissed the site's contents as being too mixed up to be useful.[10]

Another hurdle in proving that the mammalian extinctions were caused by humans is that there is no direct evidence that humans over-hunted Australia's strange giants. Australian Aborigines did not start manufacturing stone tools until thousands of years after the megafauna had vanished. They must have hunted during their first years in the new land but did so with simple wooden spears, long turned to dust. The idea that people drove the extinctions is based on inference, on the matchup in timing between human arrival and the end of the megafauna's long hey-day, and on the global pattern of a deadly syncopation between the expansion of human populations and the disappearance of native mammals.

For Chris Johnson, an ecologist at James Cook University in Queensland, the contrast between the lost and the survivors is revealing. Big, slow-breeding creatures that lived in open habitats disappeared, while smaller species that turned out offspring quickly, or lived hidden in treetops or dense forests, survived. That supports the idea that human predators pushed the big marsupials over the edge. Johnson doesn't imagine the process as having been particularly swift or dramatic— he envisions a slow process of attrition, during which people occasionally killed a giant kangaroo or a *Diprotodon* when the opportunity arose.

To explore this concept, Johnson, along with population biologist Barry Brook, created a computer model to simulate the fate of *Diprotodon* in a land newly colonized by humans.[11] No one knows the detailed life history of this long-extinct creature, but the two researchers based their estimates on data from living large marsupials and extrapolated the numbers to fit what is known about *Diprotodon's* large body size and feeding patterns. Like a modern kangaroo, the ancient giant would have birthed only one offspring at a time, and each would have taken years to grow to sexual maturity. A five-year-old *Diprotodon* might have been about half the size of an average adult, and it would have been a much easier target for a hunter. In Johnson and Brook's model, a clan of ten people could wipe out a local *Diprotodon* population by killing only two animals per year. The levels of hunting needed to drive the species to extinction might have represented only a small investment of time by the hunters and made only a minor contribution to their food supply.

Johnson points out another telling difference between the quick and the dead: in many cases, the survivors shrank. While the shrinkage of America's Pleistocene horses has been seen as evidence of a climatic cause for their decline, hunting pressure can also favor smaller body size among prey. In the life history of an animal, growth and reproduction normally trade off against each other. A young animal grows, stops at a certain age, and channels surplus energy into reproduction. If hunting is an important cause of mortality, then the sooner an animal reaches sexual maturity, the better its chance of successfully producing offspring before it dies.[12] The opposite strategy, longevity, is an adaptation for times of harsh and unpredictable climate. Several lines of living marsupials, including koalas, devils, kangaroos, and wombats, had giant Pleistocene ancestors. *Macropus titan*, at least twice the size of the largest living kangaroos, was at first assumed to be a distinct species but is now regarded as a large ancestor of the extant eastern gray kangaroo. The same pattern holds among echidnas, the spiky, egg-laying mammals that waddle through modern Australian savannas. Their Ice Age ancestors were much larger than the modern version.

In Australia, the world's most flammable continent, there is yet another controversial theory surrounding the fate of these extinct giants. Some scientists believe that humans did drive the megafauna to extinction, but that hunting may have had little to do with it. As they see it, people did the deed not with spears but with torches.

VISIONS OF FIRE

As the sky darkens, the outlines of the savanna trees vanish into blackness. Cyrus and Lindsay Rostron, Aboriginal boys 13 and 11 years old, hold fire sticks in their hands. They touch the burning branches to the grass, and flames blossom. Dancing barefoot in the orange glow, the boys are silhouetted against the tall, weird shapes of termite mounds. Their older brother, Miko, eases his muscular body off a rock that sits too close to the fire. He gives Lindsay an encouraging slap on the shoulder and then fans himself with his sooty bowler hat and stands back to admire the blaze.

David Bowman, a tall, wiry man with a hawkish nose, grins in the firelight, his face streaked with sweat and charcoal. Bowman, a fire ecologist at the University of Tasmania, is learning how indigenous people use fire to shape their homelands and keep the land healthy. One of the first scientists to study traditional burning practices by working directly with natives, he has spent years studying the way the Rostrons burn their home turf along the Cadell River in Arnhem Land, an Aboriginal "designated land" that stretches from the edge of the Arafura Sea deep into the stark landscape of Australia's Top End.

The Rostrons are among the few Aboriginal families who have been able to carry on their burning traditions without interruption. In October, at the close of the dry season, the grasses on their land are desiccated tinder, and lightning storms have begun to march across the savanna. Yet they light their fires in a casual, joyous way. That night on the Cadell River, Lindsay strolled among flaming blades of grass, playing with Bowman's digital camera. The fire moved leisurely through the dry grass, never reaching high enough to scorch the leaves of the eucalyptus trees. It would fizzle out on its own by morning. That stands in stark contrast to conditions in most of Australia, where bushfires rage with growing intensity in the absence of native management. An understanding of the way Aboriginal people use fire could be vital to the country's future.

Near the tiny Arnhem Land settlements or "outstations" that serve as home bases for native people who have traded town life for residence on traditional land, Aborigines still burn on a regular basis, and wildlife thrives. Plant and animal diversity is as great as, or greater than, in some ecologically similar areas of nearby Kakadu National Park, a World Heritage conservation site. Within weeks of the fires set by outstation residents, scorched areas sprout carpets of new greenery and attract kangaroos, which fatten on the new growth and can later be hunted. Species such as the partridge pigeon and the northern quoll, a meat-eating marsupial about the size of a house cat, have declined or disappeared throughout most of their range but remain abundant near the settlements. Wild fruit trees flourish. The Bininj, as the native people here call themselves, accomplish all this in ways that most white, or Balanda, fire managers believe to be impossible: by burning at the height of the dry season, guided only by long experience and an intimate knowledge of their home patch.

With or without Aboriginal management, Australian landscapes will burn. Adapted to lightning and dry heat, many native plants need fire and cannot reproduce without it. Most eucalyptus trees, for example, drop seeds only after being scorched, and their seedlings grow best in a bed of ashes.[1] Some species have evolved effective ways of spreading flames; for example, the candlebark eucalyptus, when ignited, drops flakes of burning bark that float on the wind and can start new fires several miles away.

During Australian summers, media attention usually focuses on bushfires in heavily populated parts of the country. In January 2003, for example, lightning in national parks outside of Canberra touched off blazes that roared into the city, destroying more than 500 homes and killing four people. That same season, nearly 94 million acres burned in the northern and central sections of the continent. The impact on people and property in these remote regions may seem relatively minor.

Yet bushfires are growing bigger and hotter, a change that bodes ill for many native animals and plants.

Before Europeans arrived, Aboriginal people set frequent small fires. They used flames to fireproof the landscape, a paradox many Balandas find difficult to grasp. When the natives lived in small groups scattered throughout the Top End, they started blazes so often that the flames seldom spread far or grew too hot. The result was a patchwork of firebreaks, areas cleared of flammable grass and trees. Fires would peter out at the edges of previously burned areas, so that even in the scorching heat at the end of the dry season, they rarely flared out of control.

With the arrival of white settlers, many native people died from introduced diseases or in violent confrontations with the newcomers. Ancient Aboriginal societies across the continent collapsed. The Northern Territory was one of the last regions affected. Arnhem Land, which covers more than 37,000 square miles, was once home to clans of Aborigines who lived scattered at regular intervals across the landscape. Today that wide swath of country is mostly empty. Arnhem Land's total population is about 17,000, two-thirds of it indigenous, and most of the people live in overcrowded government-issue houses in towns that have sprung up in the years since colonization.

Bowman was drawn to work with Aboriginal people when he noticed that their outstations were surrounded by thriving stands of cypress pine, a fire-sensitive tree that is dying off across much of its range. "If cypress pines were people," he says, "we'd talk about an epidemic. If you go to nearby places where no Aboriginal people have been living for 50 years or so, you can fly along in a helicopter and count dead cypress pines. They're just scattered all over the landscape." In the Rostron's country on the Cadell River, however, he found an abundance of healthy cypress pines, evidence of the Aborigines' skillful use of fire. He would eventually find that native species, from kangaroos to fruit trees, all flourished under native management, while they dwindled away in abandoned country where fire starting was left to random strikes of lightning or the uncoordinated efforts of Balanda landowners.[2]

Bowman calls the situation in Arnhem Land a "research emergency." Ancient land management traditions are as risk of disappearing, as fewer native people remain on their country. Federal agencies have supplied funding for a project aimed at helping to sustain native burning practices and passing them on to the next generation. Yet such programs often impose Balanda ideas about the proper timing and distribution of fire, warping the patterns of flame and regrowth in the few places where Aboriginal management has survived.

To study fire in Arnhem Land, Bowman must combine the high-tech tools of modern science with a profound respect for native culture. In Aboriginal societies, direct questions on such personal topics as the way a man burns the land can seem rude. The response can range from silence to long discourses on apparently unrelated topics, such as the nature of cloud formations in monsoon time or the mating habits of kangaroos. Bowman has succeeded in crossing a vast cultural divide not only because of his passion for fire ecology, but also because he has come to care deeply for the native people who have taught him so much.

In 1999, a paper published in *Science*, and widely covered in the popular media, claimed that the ancient forebears of modern Aborigines had altered whole ecosystems with fire, spelling doom for the continent's biggest beasts.[3] Bowman promptly clashed with Gifford Miller, a climate scientist at the University of Colorado–Boulder and lead author of the *Science* paper. He feared that Miller's theory would be used to discredit Aboriginal management practices, which he sees as crucial to the conservation of modern Australian landscapes.

An affable man who tends to think, speak, and act at high speed, Bowman did more than gripe about Miller's vision of a human-caused fire catastrophe in Pleistocene Australia. He began exploring the ecology of the vanished megafauna, an area of research he had not focused on before. Over the next several years, he ventured into the volatile scientific argument about the cause of the extinctions. In the end, he concluded that humans had indeed triggered the demise of Australia's giants. "The question is no longer if, but rather how, humans induced this prehistoric extinction event," he wrote in 2007.[4]

Miller, for his part, had set out to understand ancient climates, not megafauna. He had come to central Australia to study the waxing and waning of the monsoon, a powerful storm system that throughout much of the continent's past carried seasonal rains far into the interior. The monsoon grew stronger during Pleistocene warm spells, fading away when glaciers grew to cover much of the northern hemisphere. The known global

Figs. 18 and 19 Some evidence suggests that Australia's first human settlers contributed to the continent's mass megafauna extinction by burning the landscape, triggering a major shift in vegetation more than 40,000 years ago. Fire ecologist David Bowman (top) has shown that modern Aboriginal fire management maintains biodiversity and keeps threatened plants and animals flourishing. Tom and Cyrus Rostron (bottom) routinely burn their land in a remote stretch of the Northern Territory. (Photos by Hugh Scanlon.)

pattern suggested that at the onset of the Holocene, 10,000 years bp, the monsoon should have returned. It came back to most of the planet's tropical regions, in the Americas, Asia, and Africa. But for some reason the monsoon abandoned post-Pleistocene Australia. Monsoonal rains still visit the Top End, at the continent's northernmost edge. But the interior remains arid, and Miller wanted to understand why.

An ideal place to examine this question is the Lake Eyre basin, which covers one-sixth of the continent, a vast landscape that drains into a single salt lake that lies 65 feet below sea level. Today, this area is a bleak salt playa, but it once held a lake almost 100 feet deep, twitching with the movement of crocodiles and surrounded by lush wetland. Miller and other geologists could read the record of changing wet and dry habitats in layers of sediment left behind in the basin, and this offered a remarkable logbook of shifting Ice Age climates. The problem was finding a reliable way to date the different layers. Miller turned to fossil egg shells he found in dune deposits along the old shoreline of the vanished lake.

The eggshells had been left by ancient emus, ostrich-like flightless birds that still walk Australia's savannas and woodlands, and by *Genyornis newtoni*, an extinct bird species whose massive bones suggest each individual would have weighed about 550 pounds, or five times more than an emu. Some of the shells had been scorched in long-ago fires. Miller, who had learned how to use ancient eggshells to read the passage of time while working on a previous project in Africa, saw them as the key to sorting out the chronology of Lake Eyre's climate.

Australia's acid soils and severe climate quickly leach all organic matter out of bone. But because eggshells have a different mineral structure than bone, the ancient shells retained traces of protein. This made it possible to date them using a technique called amino acid racemization (AAR), which is reliable over a much longer time span than radiocarbon dating. When animals synthesize protein, their bodies use only the left-handed isomer of the amino acid isoleucine. (Organic molecules occur as right- or left-handed isomers. These contain the same atomic structure but are mirror images of each other.) After death, the amino acid molecules begin to revert to the right-handed form, ultimately ending with an equal number of each type of isomer. The AAR technique uses the ratio of left-handed to right-handed isomers to determine the age of a fossil. Results from a large set of samples—1,200 dates collected from three different sites—showed that the emu and *Genyornis* had coexisted for millennia. Then, about 45,000–50,000 years ago, *Genyornis* vanished. Miller's group became the first to succeed in directly dating remains of Australia's lost megafauna. Tim Flannery, a paleontologist at Macquarie University and a prolific writer who had popularized the idea of human overkill in his 1994 book *The Future Eaters*, called the finding "a major breakthrough in Australian prehistory."

The eggshells also offered clues to the big birds' diets. Grasses that thrive in hot habitats use a unique chemical pathway to capture the sun's energy, a process that skews the amount of the stable isotope carbon-13 they contain relative to most other plants. Miller and his colleagues compared the carbon isotope signatures of fossil emu and *Genyornis* eggshells. The results show that the two bird species relied on different food sources. Grass must have been abundant right before the extinction event, because the emu of that era ate little else. The doomed *Genyornis* ate both shrubs and grass.

Two large, flightless birds living in the same habitat would have to develop different survival strategies, and different food sources, or constantly clash with each other. The emu is now, as it was during the Pleistocene, a flexible eater. If necessary, it ate only grass, but it could also make do with shrubs and herbs. *Genyornis* included shrubs in its menu but seems to have been unable to get by without grass. About 45,000 years bp, the vegetation changed throughout the Lake Eyre basin. Emus shifted their tactics and began eating lots of shrubs. *Genyornis* could not adapt, and the species died out. More recently, Miller has analyzed the carbon isotope signatures of fossil wombat teeth from his study sites. Like the emu, the wombat survived the ecological shift that came 45,000 years ago because it fed on shrubs, making do without grass.[5]

According to Miller, the arrival of people, carrying fire sticks in their hands, changed everything. The harsh landscape of interior Australia does not hold well-preserved deposits of ancient pollen that might precisely track the changing vegetation. Based on the shifting diets of the herbivores he has studied, however, Miller can envision how and why the balance tipped. "Before people came, there must have been a savanna mixed with open woodland, the kind of habitat where grasses are abundant in years of good rainfall and the trees and shrubs sustain the animals when water is scarce," he explains. Frequent burning tends to favor grasses over shrubs and trees, yet at the moment of human arrival, emus and wombats stopped eating grass, and *Genyornis* simply vanished. Miller argues that burning did promote grass growth, but that grass was spinifex, an unpalatable species that now dominates the region. Spinifex leaves are heavily loaded with silica and nutrient-poor. No native mammal eats the stuff, and introduced cattle will take only the freshest, youngest shoots that sprout up after a fire. And spinifex loves fire: it grows in hummocks where live and dead blades intermix, laden with flammable resin, and it resprouts quickly after a burn.

Australian plants had evolved with fire for millions of years. Before people came, the bush burned regularly at the end of the dry season, the time of many lightning strikes. But humans could light fires at any time of the year. In Miller's scenario, the increased frequency of fire wiped out many plants. Many ecosystems in the interior, he says, are adapted to burn every twenty to fifty years, the kind of frequency that occurs if lightning is the only ignition source. Torched frequently by Aboriginal people, such habitats would not have enough time to recover between burns.

Miller theorized that human-lit fires broke up big stands of mulga, a species of acacia tree that now grows only in scattered patches, surrounded by swaths of spinifex. He suggests that this change was so drastic that it caused the failure of the monsoon. A powerful feedback exists between vegetation and rainfall. Plants shape climate by recycling water and releasing it back into the atmosphere. If Miller is right about this—though he admits it is very difficult to prove—then humans were altering the weather long before we burned fossil fuels.

Seeking his own answers, Bowman went into the Tanami Desert in central Australia to study the fire ecologies of mulga and spinifex. Traditional Aboriginal burning in the region had ceased fifty years earlier, but by reading isotopic evidence from soils he was able to track the distribution of trees and grasses more than a thousand years back in time. He found himself agreeing with Miller.[6] Mulga woodlands had likely been more widespread early in the Pleistocene. The arrival of Aborigines increased fire frequencies, favoring the rapid expansion of spinifex and the shrinking of woodlands. As the megafauna died out, more uneaten fuel was left to burn, creating a cycle of intense fire that favored spinifex. By the time

Europeans arrived on the scene, however, Aboriginal people had become masterful fire managers. In the Tanami, as in Arnhem Land, they created a mosaic of burned and unburned patches, a landscape where fire would not flare out of control. That's a marked contrast to the current reality, in which fires burn over vast areas in a single season. As a result, fire-adapted creatures such as the western hare-wallaby, which had long coexisted with native people, are slowly disappearing from the region.

Working with population biologist Barry Brook, Bowman helped construct ecological models showing that big, slow-breeding animals would have been particularly susceptible to extinction upon human invasion.[7] That part of the story, they point out, is not over. Many of the mammals that survived the Pleistocene extinction crisis are now in decline, hard-hit by accelerating human impacts on everything from water supplies to open lands to weather. Their ancestors may have helped wipe out *Diprotodon* and *Thylacoleo*, but the fire practices of Aboriginal people may be the key to rescuing dwindling populations of quolls, bettongs, and wallabies.

No longer bothered by the idea that Australia's first people created major changes in their new environment, Bowman now believes the mass extinctions were caused by an ensemble of factors, including human-driven changes in fire regimes, hunting pressure, and natural climate change. "I don't give a stuff whether they killed the megafauna," he says. "I care about Aboriginal people, and I know that they achieved a sustainable balance. It's tragic that we've underappreciated the contribution native people have made in creating biologically diverse landscapes."

Years after Miller published his first, controversial paper on vanished Australian megafauna, he and Bowman agree on both the powerful impact of the continent's first people and the importance of traditional Aboriginal management. "The continent's remaining biodiversity can only be maintained if we continue Aboriginal fire practices," says Miller. "All of Australia is a man-made ecosystem, and has been for tens of thousands of years." Some native people embrace Miller's vision of their ancestors as powerful folk, capable of driving giant beasts to extinction. The idea is a welcome antidote to the old British colonial claim that Aborigines had never altered their environment and so had no rights to the land.

Soon after the publication of his 1999 paper, pinning the blame for megafauna extinctions on the continent's first people, Miller traveled to remote northwest Australia to continue his research. He was guided by a group of aged Aboriginal ladies. They found the area around his sampling site blanketed in the grass that comes with monsoon rains, a dense green mass that stands taller than a man's head. "The women said, 'Burn this! We won't walk through this, there's snakes in there,'" recalls Miller. Fire, so feared by Balandas, is a basic instinct for the Bininj.

The first people in Australia reshaped the landscape, then learned, in intimate detail, how to maintain the plants and animals that had survived the initial shake-up. Modern land managers can ignore Aboriginal knowledge, but in doing so risk losing much of the biodiversity that remains. Today we are repeating the unplanned experiment conducted by ancient Aboriginal people but on a global, rather than a continental, scale. Modern humanity is blindly altering the planet—through pollution and habitat destruction—without knowing what the consequences will be.

This is perhaps more obvious in Australia than in many other parts of the world. The massive changes brought by Europeans have created turmoil in the ecosystems that Aboriginal people knew and managed for millennia. Even before they began to quash ancient fire management systems, Balandas brought hoofed grazing animals: cattle, sheep, horses, camels. Some of these creatures have formed the basis of major industries that have transformed woodlands and wild savanna into pasture. Others have gone feral, thriving in the ecological gaps left by the demise of the native megafauna. Heedless introductions of house cats, foxes, and rabbits resulted in a deadly triumvirate that hunt and outcompete small native marsupials. Over the past 200 years, nearly half of all mammalian extinctions worldwide have occurred in Australia. Eighteen endemic species, most of them common and widespread at the time of European contact, have now disappeared forever; another nine survive as small remnant populations in islands of protected habitat.[8] The ongoing small mammal extinctions, says Bowman, form a mirror image of the Pleistocene megafauna die-off, which likely began in the Top End where the first humans landed. Small mammal populations first collapsed in the center of the continent, and a wave of extinctions continues to sweep north.

One sweltering October day in Arnhem Land, Bowman drove the men of the Rostron family over a bone-rattling track through open woodland. Miko carried a rifle of World War I vintage, and twice they stopped as he tried to take down a water buffalo. The first time, he couldn't get close enough, and he held his fire, saving precious ammunition. The second time they came close to a skinny cow who waved her horns lazily in the late afternoon sun. Miko crept nearer and shot, but the buffalo, only wounded, ran off.

Water buffalo, native to Asia, were introduced to the Northern Territory in the early 1800s to serve as beasts of burden for early British outposts. When those settlements collapsed, the buffalo went feral and thrived. Today they are a normal part of life in Aboriginal communities where poverty is endemic and food is often scarce. The Bininj call these animals Nganabbarru, and they are such skillful hunters that even with a few antiquated guns they manage to harvest a fair amount of meat. Indigenous people would rather eat emu or kangaroo, but these creatures are savvy and fleet of foot: catching them most often requires an organized hunt involving many people. The right to kill native wildlife is also entwined in complex clan rules of land ownership and responsibility, a code that does not encompass imports like the buffalo.

Many Balandas now consider the feral buffalo a scourge. The animals carry brucellosis and bovine tuberculosis, deadly diseases that can infect domestic cattle. During the 1980s, Australia's government spent hundreds of millions of dollars trying to eradicate feral buffalo. Even with trained hunters shooting from helicopters, the effort failed. Buffalo numbers were knocked down, but killing off the last few members of remote populations proved impossible. The fewer animals there are, the faster the survivors breed. Besides, the rarer the buffalo become, the harder they are to find at all in the vast wild sweep of the Northern Territory.

It's easy to argue that the buffalo have no place in Arnhem Land. Any hoofed creature is foreign to Australia, and feral cattle stomp and churn wetland mud and forest soil, affecting the habitat of threatened native creatures. Still, the buffalo's ability to thrive there implies it is, in some ways, filling an environmental slot left empty by the disappearance of *Diprotodon* and the short-faced kangaroos—even introduced bovines can shed light on the causes of the Pleistocene extinctions. The water buffalo's successful invasion of Arnhem Land has been possible because it is an adaptable eater, feeding on grass during the wet season and switching to shrubby browse during dry times. The banteng, an endangered species of wild cattle from Southeast Asia, was also introduced to northern Australia about the same time as the water buffalo, but its population remains limited to a small area near its original point of release. The banteng is a browser, unable to survive on grass. Like the vanished marsupial giants of the Pleistocene, the banteng is poorly adapted to a landscape dominated by fire-wielding people and the grasslands they create.[9]

A decade ago, Bowman actively worried about the claim that Australia's first people were culpable in the megafauna extinctions. He feared that white society would use the excuse to write off the value of Aboriginal land management. Today, the politics of living and dead megafauna have shifted. The idea that Aboriginal people might be able to exterminate large creatures sounds appealing to white Australians who want to control populations of buffalo and other feral animals. Returning the Bininj to their traditional lands appears healthy both for the environment and the people. (Aboriginal people living in Balanda-built townships suffer from high rates of unemployment, diabetes, obesity, and cardiovascular disease. A recent study found that individuals who live on and manage their lands get more exercise, have a better diet, and are healthier mentally and physically than their town counterparts.)[10]

"It's all very complicated," muses Bowman. "Aborigines had an effect. They contributed to the extinctions, but then they stabilized the system and carried a whole lot of species through with them to the Holocene. We came along, and the whole damn thing just fell apart." That pithy description sums up the interaction between European colonists, indigenous people, and wild ecosystems around the world. Understanding the fall of the Pleistocene megafauna can help define ways to repair more recent damage.

In parts of the western United States, biologists have slowly begun to replace some species of missing megafauna, vital cogs in the dented ecological machine. The revival of the wolf in Yellowstone National Park, a hard-fought conservation battle, has succeeded in surprising and impressive ways. Lovers of the lost Pleistocene world now hope to leverage that small first step into a great leap away from long-standing conservation dogma.

WILD
DREAMS

THE WOLF RETURNS

Fleas still hop at the den entrance, dug into the side of a small hill on the rolling surface of the Blacktail Deer Plateau in Yellowstone National Park. It's early September, and a wolf pack with eight adolescent pups abandoned this safe haven within the last ten days. The predators left plenty behind—scats, a scattering of sticks, and elk bones marked with the fine scratches of sharp puppy teeth. Still, walk a few steps away and the den site vanishes into the undulating plain, camouflaged amid the wild rose and sweet-smelling sage.

Like its den, the gray wolf's impact on Yellowstone, though profound, can be difficult to spot with the untrained eye. Extirpated in the region in the 1920s, the species was reintroduced to the national park in 1995. Since then, scientists have uncovered evidence that when the wolf vanished from the park, other species were also powerfully affected. Wolves shape the destinies not only of their prey and rival predators but also of plants and songbirds. "With the wolf back in place as the top carnivore, biodiversity is greater," says biologist Doug Smith, leader of the Yellowstone Wolf Project. "The return of the wolf is the best thing to happen to Yellowstone in the past century." The drama playing out in Yellowstone can be seen as a Pleistocene parable, an object lesson in the importance of megafauna. The earth needs big beasts, both herbivores and predators. Each group relies on the other, in ways we are only beginning to understand.

Smith, a tall, lean man who has been tracking the lives of Yellowstone's wolves since the first animals were imported from Canada in 1995, walks downslope from the den. There, by a small pool of water, lies an elk carcass.

The mud around it holds tracks of both wolf and bear: grizzlies often scavenge at wolf kills. Every scrap of meat has been devoured, but the spine is intact, and a patch of thick, wiry, nut-brown fur still covers the base of the antlers. Smith pulls a tooth out of the elk's jaw, which he will use to determine the animal's age. By cracking open the long bones, he can learn the elk's state of health when it was killed: in a strong animal, the fat-rich marrow coheres like a Jell-O mold; an animal weakened by hunger or sickness will begin digesting its internal fat to survive, and the marrow turns to runny, amorphous slime.

The relationship between wolves and elk is key to understanding the sweeping changes that have come to Yellowstone in the past 150 years, since Euro-Americans discovered the area in the 1860s. Wildlife was hunted intensively at first and then protected piecemeal according to prevailing attitudes. Elk, targeted by commercial hunters in the nineteenth century, were later put under protection, while park managers mandated destruction of predators such as wolves and cougars. Today elk are plentiful, and populations of wolves, as well as mountain lions and black and grizzly bears, are recovering.

These extreme shifts in both predator and prey populations have changed not only the number and kinds of megafauna in the park but also the growth of aspen, cottonwood, and willow, the survival of beavers and birds, and even the shape of streambeds. When wolves disappeared early in the twentieth century, their loss set off a cascade of ecological change.

In photographs taken in the 1890s, aspen groves stand tall on Blacktail Deer Plateau. Many of those trees are now gone. Since Yellowstone's establishment as a national park in 1872, most of its aspen forests have been lost.[1] Though they never covered more than 4 percent of the park's northern range, aspen groves support a greater variety and abundance of birds and understory plants than do surrounding Douglas fir and lodgepole pine forests.

Soon after the wolves returned, ecologist William Ripple of Oregon State University came to Yellowstone to study the dwindling aspen. He and his coworkers discovered that the history of the park's elk was recorded in the growth rings of surviving trees. Aspen that managed to grow to tree height—having escaped being eaten by elk as young sprouts—had all begun their lives between 1700 and the 1920s. After that elk numbers surged, and aspen seldom managed to grow taller than an elk can reach – about 6.5 feet.[2] Elk also devoured young streamside willow and cottonwood.

Fifteen years after the wolf's renaissance, aspen, willow, and cottonwood grow tall once again inside wolf territories. Even before the predators had significantly lowered their numbers, fear affected the elk's behavior, and they foraged less in areas where the risk of wolf attack was high. By 2005, wolf packs had claimed all the usable habitat in the park, and elk numbers began to drop. The dramatic changes Ripple and his colleague, Robert Beschta, have documented result from both the decline in the elk population and shifts in the animals' behavior.

Wolves tend to travel along streams, and it is there that the regrowth is most obvious. By the spring of 2006, Ripple and Beschta found that willow shoots in riparian areas had grown taller than a feeding elk could reach—this had not happened on the northern range for more than half a century.[3] Aspen and cottonwood are also stretching toward the sky, especially on creek flats, where wolves are most successful in taking down their prey.[4]

In the years when the elk reigned, unchecked by top predators, other species suffered from the suppression of riparian plants. Beavers had flourished on the northern range in the 1920s, but their population dwindled as aspen, willow, and cottonwood groves were overbrowsed by elk. By the 1950s, streamside aspen had all but disappeared, and surviving willow and cottonwood had grown thick, heavy stems. Beaver numbers plummeted. In 1985, beaver were reintroduced in the Absaroka-Beartooth Wilderness, on U.S. Forest Service land north of Yellowstone. A few of these animals worked their way downstream onto the park's northern range.[5] Now, the beaver are booming in areas frequented by wolves, which once again yield good forage and dam-building material. Active beaver colonies on the northern range increased from one in 1996 to ten in 2005, and Smith thinks the big rodents have reached their carrying capacity. They now subsist on willow, which regenerates faster than aspen and thrives around beaver lodges. The wolf, the beaver, and riparian trees have powerful impacts on one another. Wolves often prey on beaver; beaver need young aspen and willow for food and dams, and wolves control elk, allowing the plants to grow; beaver dams flood the land, creating more habitat for the trees.

The fate of beaver in Banff National Park in Canada also appears linked to that of wolves and elk. Elk numbers boomed after wolves were extirpated from the area in the 1950s. According to Mark Hebblewhite, an ecologist at the University of Montana who did his doctoral research in Banff, elk and beaver competed for the available growth of willow until wolves recolonized Banff's Bow Valley in 1987. Cliff Nietvelt, another Banff researcher, found that beaver select willow stems of a size class that vanishes when elk populations are dense.[6] In Banff, which is a much harsher habitat for elk than Yellowstone, the return of the wolf reduced the elk population by 50–60 percent, allowing a rapid rebirth of willow and aspen in wolf territories.

Beaver dams create ponds that are biodiversity hotspots, hosting many insects, fish, birds, and amphibians that prefer still water. Hebblewhite remembers the only time he saw boreal toads breeding in Banff, at a beaver pond 1.25 miles from an active wolf den. The pond was surrounded by abundant willow, in an area that had been overbrowsed during the years when wolves were absent from the park. Some researchers believe beaver ponds are critical habitat for the boreal chorus frog, another rare amphibian now being studied in Yellowstone.

South of Yellowstone, Grand Teton National Park, along with the nearby national forest lands surrounding Jackson Hole, Wyoming, remain nearly empty of big predators like wolves and grizzlies. Here the problem is moose, the largest living member of the deer family and a browser with an insatiable appetite for willow. Moose were rare around Grand Teton until the 1880s, when people began industriously killing off wolves and grizzlies. In the absence of these top predators, the moose population exploded. Today it is common to see groups of thirty or more moose devouring the scrawny willows inside the park. Moose densities inside Grand Teton are about five times higher than outside the park, where moose hunting is allowed.

Inside the park, browsing by an overabundant moose population leaves willows stunted, studded with dead stems. Songbirds are less diverse and fewer than in healthier willow stands outside the park. Two species that depend on dense thickets of willow for nesting habitat, MacGillivray's warbler and the gray catbird, are absent from the park but are found outside it.[7]

The discovery that bird diversity was higher outside the national park came as a surprise, notes Peter Stacey, a biologist at University of

New Mexico who conducted the bird counts. Parks are refuges for wild-life, and most people assume that the absence of hunting increases biodiversity. The researchers realized that the reality was the opposite of that expectation—because Grand Teton is missing a key ecosystem component: an effective top predator.

In Banff, Hebblewhite and Nietvelt found that in areas where wolves had reduced elk density, thriving willow habitats hosted a diverse array of songbirds—including the American redstart, a species that needs mature willow to nest successfully. Near the town of Banff, where elk had taken up residence on lawns and golf courses, safe from wolf predation, willows did not grow above ankle height, and redstarts didn't nest at all. These data revealed a direct relationship between wolf density, elk density, the regeneration of aspen and willow, and songbird abundance and diversity.[8] In the span of 3 miles from the town of Banff, changes in the distribution of all these species are visible.

Wolves avoid people as much as they can, so when they recolonized Banff's Bow Valley, elk fled to the safety of the town. "All of a sudden," says Hebblewhite, "elk were living on people's front lawns, and never seeing anything scarier than a Shih Tzu." Several residents were injured in elk encounters, and a committee of scientists, Hebblewhite among them, was asked to solve Banff's urban elk problem. Their strategy: restore corridors of wolf habitat leading into town, bringing the predators to the prey. The plan appears to be working, reducing the number of elk and redistributing the animals that remain.

The presence of wolves echoes in powerful ways through the entire ecosystem, a phenomenon ecologists call a "trophic cascade." Researchers have now documented parallel cases in diverse habitats worldwide, in which the loss of an apex predator has had profound repercussions. The extirpation of sea otters along much of North America's Pacific coast has led to the destruction of kelp forests, which are now overgrazed by sea urchins.[9] A famous study of predator-free islands, isolated when a dam flooded the Caroni Valley in Venezuela, showed that growth of young trees plummeted and entire ecosystems were devastated in the absence of large carnivores.[10] In much of eastern North America, where wolves and cougars were wiped out in the 1800s, populations of deer have grown to plague proportions, destroying native plants through overbrowsing, stunting tree growth, and spreading Lyme disease, which is carried by deer ticks.[11]

The first wolf pack to settle on Yellowstone's Blacktail Deer Plateau after the 1995 reintroduction was named after Aldo Leopold, a man who recognized trophic cascades before anyone had given them a name. He is remembered as an early crusader for carnivore conservation. Yet he began his career as a wildlife biologist enthusiastically hunting down wolves and cougars—a popular tactic in the days when wildlife management meant maximizing the number of huntable deer and elk. Leopold described the way the death of one wolf changed his mind. "We reached the old wolf in time to watch a fierce green fire dying in her eyes," he wrote in his popular essay, "Thinking Like a Mountain." "I thought that because fewer wolves meant more deer, that no wolves would mean hunter's paradise. But after seeing the green fire die, I sensed that neither the wolf nor the mountain agreed with such a view."[12] Before many now-prominent wolf researchers were born, Leopold began to advocate for the return of wolves to Yellowstone and other "rugged mountain places." "The Yellowstone has lost its wolves and cougars, with the result that elk are ruining the flora, particularly on the winter range," he noted.[13]

Having witnessed the intentional extirpation of wolves and cougars, Leopold was among the first to understand the consequences. By the 1930s, mass irruptions of deer were taking a heavy toll on significant swaths of the west, halting the growth of trees and shrubs. Leopold made the connection between predator loss and herbivore explosions, a widespread pattern he documented most famously on the Kaibab Plateau in Arizona.[14] When the Kaibab's high-elevation forests and meadows were declared a game reserve in 1906, deer were protected while predators were pursued relentlessly by government-hired hunters. By 1931, hundreds of mountain lions and bobcats, thousands of coyotes, and the last of the local wolves had been killed. Meanwhile, the deer population, about 4,000 strong in 1906, mushroomed to an estimated 100,000 in 1924, bringing the growth of young aspen to a halt.[15] Thousands of deer died of starvation in the mid-1920s. The deer boom degraded habitat and lowered the area's carrying capacity for game. Leopold's Kaibab study was long considered a classic, taught to every student of ecology and wildlife biology. Later, in the 1970s, competing theories on the causes of deer irruptions—pinning the blame on other factors, including fire suppression and livestock grazing—became more popular, and the Kaibab example vanished from textbooks.

Fired up by their dramatic findings in Yellowstone, Ripple and Beschta have studied five other western national parks, seeking evidence of altered trophic cascades where top predators have vanished. They found that from Jasper National Park in Canada to Yosemite in California, when large predators dwindled, hardwood trees followed suit. In Olympic National Park in Washington, few new bigleaf maple or cottonwood trees have been able to grow since wolves were eradicated in 1910. Elk were freed to browse without limits and have transformed riverbanks from the thickets of dense brush described by early explorers to open, parklike stands of big trees. Without the roots of young saplings to hold stream banks in place, erosion has increased. Rivers that were held in tight channels during the days of the wolf have widened, hosting fewer fish, while plants and insects become scarcer along their banks.[16]

In Zion National Park, Utah, only 62 miles northwest of Leopold's study site on the Kaibab Plateau, Ripple and Beschta have documented the differences between two adjacent canyons, one with cougars and one without. When Zion Canyon was designated a national park in the 1920s, the number of people using the area skyrocketed, from 8,400 visitors in 1924 to more than 68,000 in 1934. Cougars, which go out of their way to avoid contact with people, abandoned the canyon but continued to thrive in the nearby North Creek drainage. As human use increased and cougars faded away in the 1930s, the deer population in Zion Canyon boomed from less than 80 to about 600 at its peak in 1942. Today Zion, still heavily used by people and deer, poses a stark contrast to neighboring North Creek. Only a few scattered new cottonwoods have grown up over the past ninety years, on steep slopes inaccessible to deer.[17] Stream banks are much more eroded than those along North Creek; living things of all sorts, including riparian wildflowers, amphibians, lizards, fish, and butterflies, are more numerous and diverse where cougars roam.

The Zion study serves to replicate and confirm Leopold's decades-old findings from the Kaibab Plateau. Ripple and Beschta have documented a striking pattern: when top predators decline or disappear, herbivore numbers boom, tree growth declines, and whole ecosystems can shift into an impoverished, oversimplified state. "We think this is a widespread, universal phenomenon," says Ripple. The impacts of predator loss, he suggests, may be as significant as those of climate change.

Opponents of wolf reintroduction have worried that the predators would devastate the thriving elk population on Yellowstone's northern range. But wolves in the park—as elsewhere—target the most vulnerable prey, the sick, the weak, the young, and the old. In the first few years after the wolves returned, the elk population held steady, and sometimes even increased. Many factors affect the numbers and health of elk: winter weather, rain or drought, the depredations not only of wolves but of grizzlies, cougars, and humans who are permitted to hunt animals that cross the border into neighboring Gallatin National Forest. As wolf populations have grown and spread through the Yellowstone ecosystem, elk numbers on the park's northern range have begun to decline by an average of 8 percent annually. Smith, like other wolf biologists, sees this all as part of restoring a healthy relationship between native predators, prey, and plants.

Tracking radio-collared wolves on foot and by helicopter, Smith and his colleagues have gained an intimate knowledge of when and how the wolves hunt elk, which make up more than 90 percent of their diet. An adult elk is far bigger and heavier than a wolf, and making a kill is risky business: it is not unusual for a wolf to die of injuries caused by a well-aimed hoof or antler. If an elk, or a bison, stands its ground, wolves stand little chance of prevailing. This explains why they are relatively inefficient hunters, succeeding in only about one out of five attempts to kill prey.

The locations of kills during the first decade of wolf recovery in Yellowstone show that much of the northern range consists of places where elk are relatively safe. Wolves seldom take down prey on hills or ridges. Only on flat grasslands close to streams or roads do they have much success. Such turf is so valuable that separate wolf packs, which usually enforce territorial boundaries using deadly force against their rivals, seem willing to share it.[18]

The pattern written in bloody snow at winter wolf kills matches the most recent findings from Ripple and Beschta's studies of tree growth on the northern range. In areas of high predation risk—near stream channels and in spots where fallen logs make it difficult to flee—elk no longer tarry, and aspen, cottonwood, and willow are left to grow tall. On the

Fig. 20 Wolves hunt elk in Yellowstone National Park. The predators have changed elk behavior and limited their numbers, leading to a renaissance of riparian aspen and willow that had been overbrowsed by elk during the decades when wolves were absent. (Photo courtesy of Douglas Smith, Yellowstone Wolf Project.)

uplands, however, elk browsing remains heavy, and aspen groves look much as they did in the early 1990s, before the wolves returned. The aging trees stand widely spaced, their bases scarred with the marks left by hungry elk during lean winters. Few saplings grow up in the gaps between the adult trunks.

The northern range will always hold places of refuge where elk are safe from hungry predators. These shelters are inherent in the lay of the land and in the shape, growth, and decay of shrubs and trees. In a sophisti-cated statistical analysis of wolf predation around Banff, Hebblewhite has shown that for elk, topography is a major factor affecting the risk of encountering wolves and that details of habitat (open ground on which to flee, versus obstacles such as thick brush or deep snow) often deter-mine whether an elk survives.[19]

For decades, wildlife managers, hunters, and ecologists debated the wisdom of letting abundant elk live in Yellowstone unregulated by top predators. Some have suggested that where we cannot restore wolves or cougars, human hunters must fill the gap. That strategy may work in some places, and for conservationists seeking to restore native predators it will be essential to find common ground with hunters. Still, it is important to realize that the ways of modern hunters do not duplicate those of wild carnivores.

The majority of wolf kills in 2002 were of older cow elk. Several years later, the predators shifted their focus to bull elk during early winter hunts. In each case the adult elk that fell to wolves were in poor shape, as analysis of their bone marrow shows. The new strategy of targeting bull elk was due, perhaps, to a drought in the summer and fall of 2007. Nutritious forage became scarce, and bulls, who put all their energy into battling over mates during the autumn rut, began the winter with few physical reserves.

Humans, unlike wolves, tend to seek out elk in their prime. Often these animals are pregnant cows, so that a hunter kills two elk with one shot.[20] Elk hunts in Gallatin National Forest are carefully managed, and the number of elk permits issued there has dropped since wolves returned to Yellowstone. Yet the hunters who do come to the Gallatin now move through a landscape richer and more complex than it was in the days of peak elk abundance.

Hebblewhite believes that over time the wolf will prove to have as powerful an influence on Yellowstone's ecology as it has on Banff's and that the wolf will have the last word in a long-standing argument. "Restoring predators in the areas that you can," he says, "is the best way to achieve some measure of ecological integrity. I think wolves will solve any debate there ever was about natural regulation in Yellowstone."

The gray wolf, the beast you can spot from roadsides in Yellowstone today, evolved in the Ice Age Arctic. Along with its favored prey, the elk and the bison, the wolf spread into the lower forty-eight states around 30,000 years bp, when a warm spell melted a gap in a long-standing barrier of ice. Back then the gray wolf shared territory with a menagerie of larger, heavier predators, capable of taking down mammoth and mastodon calves, as well as native horses and camels. Working with paleobiologist Blaire Van Valkenburgh, Ripple has begun to use fossil evidence to explore the trophic cascades of the Pleistocene. Their findings suggest that early human hunters in North America did trigger the megafauna extinctions—not through rapid overkill, but by tipping the balance in an ecosystem already dominated by predators.[21] Van Valkenburgh has shown that Ice Age gray wolves, along with saber-tooth cats, American lions, dire wolves and coyotes, fractured their teeth far more often than living carnivores. The state of these ancient teeth offers evidence that competition for meat was intense: predators ate every accessible scrap, and cracked bones open to get at the marrow within. Added competition from human hunters may have created enough predation pressure to push mammoths, ground sloths and other large herbivores into extinction— the opposite of the effect created when wolves were wiped out in Yellowstone in the 1920s. "We think that major ecosystem disruptions, resulting in these domino effects, can be caused either by subtracting or adding a major predator," explains Ripple. "In the case of the mammoths and saber-tooth cats, the problems may have begun by adding a predator: humans." Another kind of human-induced warp in a wild food chain is illustrated by pronghorn, graceful, antelope-like creatures that graze in and around Yellowstone. With their striped necks and elegant curved horns, they look, and act, like beasts out of the Serengeti. The pronghorn coevolved with an American version of the cheetah, a member of the now-defunct Pleistocene megafauna. Adult pronghorn can run as fast as 55 mph—a speed that makes them unreachable by any living American carnivore. The extant African cheetah, the planet's fastest land predator, is capable of sprinting at 70 mph and might be able to capture an adult. As things stand, only the fawns are vulnerable to predators of the American West. Bobcats and golden eagles will take newborns, as will wolves, if the opportunity arises. But wolves are so big that an infant pronghorn is only a snack. For a coyote, on the other hand, a fawn makes an easy feast.

In Grand Teton National Park, the pronghorn population once numbered several thousand. Intense market hunting in the late 1800s pared the herd down. Today they number only about 200 animals, the heirs to a traditional migration out of the relatively serene high pastures of Grand Teton to lower ground in the winter. When the snow flies they cross onto territory managed by the Bureau of Land Management, where numerous natural gas wells are being developed. The herd's winter range has been dug up, paved over, and packed with man-made hazards.

In spring and summer, when mother pronghorns birth their young in parts of Grand Teton and adjacent Bridger Teton National Forest,

Fig. 21   The small herd of pronghorn antelope that summer in Grand Teton National Park lost most of their newborn fawns to coyotes until wolves moved into the area from nearby Yellowstone National Park. Wolf competition with coyotes may give the herd, whose winter range is being fragmented by development, a better shot at survival. (Photo from Wikimedia Commons. http://en.wikipedia.org/wiki/File:Antilocapra_americana.jpg)

coyotes are the major threat. During the decades that wolves were gone from Yellowstone, coyote numbers boomed, reaching some of the highest densities ever recorded in North America. The top dogs changed that: wolves often attack coyotes that try to share their turf, and the coyote population in Yellowstone has been cut in half since wolves returned. But in the southern reaches of the Greater Yellowstone Ecosystem (a 30,000-square-mile swath of wildlands that includes portions of five national forests in addition to Yellowstone and Grand Teton national parks), wolves are few while coyotes abound. In a study of the Grand Teton pronghorn herd, biologist Kim Berger of the Wildlife Conservation Society found that coyotes killed nearly all of the 150 fawns born each summer.[22]

As wolves continue to spread south out of Yellowstone, they offer new hope for the pronghorns of Grand Teton. Only a single pack of wolves occupied a corner of the park during Berger's study, but inside wolf territory the survival rate for newborn fawns was four times higher than outside it. Coyote numbers in Grand Teton have dropped by a third since wolves returned, and will likely drop lower if more wolves move in from the north. The wolf effect may give the dwindling pronghorn herd a shot at long-term survival. (Many animals are still likely to die during their southbound migration, made increasingly difficult by construction of new homes and fences, or on the industrialized winter range, where precious habitat is disappearing.)[23]

The shifting balance between pronghorn and coyote in Grand Teton exemplifies an increasingly common pattern: wipe out the top predator, and smaller carnivores will boom, often with devastating and unexpected impacts. This syndrome is known in ecological lingo as "mesopredator release." In the 1990s, conservation biologist Michael Soulé coauthored a classic study that described mesopredator release in Southern California scrub habitat. The coyote was the top dog in this scenario, pushed out as native shrubs vanished under a tide of suburban development. People flooded in, building houses and fencing off manicured lawns, and coyotes faded away, leaving domestic cats as the dominant predators in the fragments of wildland that remained. The cats drove a variety of scrub-breeding bird species to local extinction: in narrow canyons too steep for building, scrub plants survive but the songs of the native birds have fallen silent.[24]

Soulé, one of the founding fathers of the young discipline of conservation biology, has spent his career finding new ways of understanding human impacts on nature. Back in the 1980s, he posited the idea that by monopolizing most of the land on the planet, *Homo sapiens* had put a halt to the normal process of evolution among large animals worldwide. Later, working with his colleague Reed Noss, Soulé proposed a broad scheme of what he called "rewilding": returning native predators to as much as possible of their original range. Only healthy habitats could sustain grizzlies and wolves, they reasoned. Restoring big carnivores was a way of returning landscapes to true wildness.

Soulé helped found the Wildlands Network, a nonprofit organization with the ambitious goal of creating and protecting corridors of habitat that would allow wildlife—especially megafauna—the space to roam on a continental scale. This idea, at first considered outlandishly idealistic, has become part of the mainstream in conservation biology. One of the earliest and best-known proposals of the Wildlands Network was a plan to keep a continuous swath of habitat open in the vast stretch of North America between Yellowstone and the Yukon, a concept known to its adherents as Y2Y.

The inspiration for Y2Y was a female wolf, dubbed Pluie by researchers who used a satellite collar to track her nomadic existence during 1991–93. In that time Pluie covered 40,000 square miles, an area fifteen times larger than Banff National Park and ten times larger than Yellowstone. Starting in Alberta, she crossed through Banff into British Columbia and then roamed into Montana, Idaho, and Washington before returning to British Columbia. The collar stopped functioning in 1993, but Pluie survived for two more years. In the end she was shot and killed by a B.C. hunter, along with her mate and their three pups.[25]

Pluie's rambling existence served to underscore a crucial point: even such seemingly vast refuges as Yellowstone won't sustain wildlife over the long term if they become lone islands of habitat in a sea of hostile, human-dominated territory. Before they died out, megafauna survived the many changes of the Pleistocene by moving freely over broad landscapes. As the world warms and human populations continue to climb, space is the one thing big wild animals need most. Giving wildlife room to move has become Soulé's great cause. "Nothing ever happens unless people push—people with vision, who are not satisfied to just stand by and watch the world deteriorate and die," he says. He has long been sympathetic to Paul Martin, purveyor of the human overkill model and passionate student of the vanished mammoths and ground sloths. In 2005, the two were among the list of coauthors on a new conservation manifesto, which appeared in the journal *Nature*.[26] The article caused a stir not only among ecologists but also in the popular press, resonating with people who live near Yellowstone and other wild places.

Along with ten other prominent scientists, Soulé and Martin proposed a radical new strategy. The loss of the Pleistocene megafauna, they reasoned, had left North America ecologically impoverished long before the coming of European settlers. Now, 13,000 years after the last mastodons died, humanity is driving many surviving species of megafauna to extinction, rapidly strangling the last gasps of wildness left on earth. Their solution: resurrect a vision of Pleistocene America. Elephants from Africa could stand in for the vanished mammoth; the endangered Bactrian camel could replace its long-dead American cousin. These giants, dwindling in the last of their native habitat, desperately need more space. For decades—long before anyone took his Pleistocene dreams seriously—Martin believed that the West could support elephants and camels, that such giant vegetarians might browse efficiently alongside cattle and even improve the ecological health of the deserts where the last ground sloths had roamed.

The return of megaherbivores would require predators of great size and power to maintain a balance. Proponents argue that Old World lions and cheetahs should be introduced to North America as well. Genetic analyses show that living African lions are smaller members of the same species that once roamed the Great Plains. The cheetah is not a direct relation of its vanished American counterpart, but the very existence of the pronghorn testifies that a niche for a high-speed predator now sits empty.

Depending whom you ask, this idea, known as "Pleistocene rewilding," represents either an inspired leap forward in the philosophy of conservation, or a burst of communal insanity. Soulé sums it up: "I'm a pragmatist. But I also think we need to push the envelope."

TALKING ABOUT A REVOLUTION

Martin had been trying to sow the seeds of Pleistocene rewilding for nearly forty years. As early as 1967, he suggested that large herbivores from other continents could—and should—fill niches left empty by the defunct giants of the American Pleistocene. He advocated for the protection of feral burros as substitutes for extinct native horses. At the time, burros were multiplying in the Grand Canyon. National park officials, concerned that introduced burros compete with native bighorn sheep for forage, wanted to kill off the intruders. Martin's early writings on this topic were dismissed by the scientific community. At a time when ecologists were documenting the devastating impacts of human-introduced exotics worldwide, the idea seemed crazy. "There is, as yet, no general agreement that people, not changing climate, were responsible for the emptying of niches," wrote Martin's mentor, Ed Deevey, in 1967. "Until the inquiry has progressed a little farther it is premature to contemplate refilling them—repopulating Arizona, say, with horses, camels, elephants, antelopes, tapirs, and peccaries—for we have one witless menagerie in New Zealand and scarcely need another."[1]

Years later, Martin found a kindred spirit when he encountered a young paleoecologist named David Burney at a scientific meeting. Burney had lived in Africa during the 1970s, where the megafauna had made him yearn for the lost giants of the American Pleistocene. He and Martin became fast friends when they realized they had been thinking along the same wildly unconventional lines: all the Ice Age animals aren't really extinct. Both men wanted to see what would happen if some of the surviving giants were returned to their ancient stomping grounds. By the

1990s, the quiet coconspirators had made a mental leap, from loving the Pleistocene to viewing it as the normal order. "What we live in is the highly abnormal world," says Burney. "We think of ground sloths and saber-toothed cats as peculiar and foreign, but it is the world of our own ancestry, the world our species evolved in." Martin's deepest faith was in the lost giants of the Pleistocene. He described himself as a "nontheist" and said that life on earth—which in his mind included the mammoth and the dire wolf—gave him more than enough to believe in.

Martin's passion for the megafauna had remained constant, but his strategy for resurrecting them evolved with the times. Early on, he argued that camels, African eland, and elephants, all shrub eaters, could be raised alongside cattle, which eat only grass. The combination would, he claimed, increase the agricultural productivity of arid western range-lands where grass is scarce. (His article ran in a 1969 issue of *Natural History* magazine under the headline "Wanted: A Suitable Herbivore to Convert 600 Million Acres of Western Scrubland to Protein.")[2] Thirty years later, in collaboration with Burney, Martin wrote a plea for what the two authors called "the ultimate in rewilding": the introduction of African or Asian elephants to the western United States.[3] They envisioned transplanted elephants roaming along the lower Colorado River or the Rio Grande, eating their way through stands of invasive Bermuda grass and tamarisk trees, which no human effort has been able to control. The elephants would also feed on the fruits of native species like mesquite—large, nutritious pods that seem perfectly designed for mega-herbivores. The piece, which appeared in the conservation magazine *Wild Earth*, waxed unabashedly sentimental, announcing that a memo-rial service for the mammoth would soon be held at the famous fossil site at Hot Springs, South Dakota.

This eloquent sermon on elephant restoration attracted little attention outside the circle of Martin's followers, but over the decades that number has grown. A few weeks after the piece ran, Harry Greene, a herpetologist and ecologist at Cornell, sat chatting with his graduate student Josh Donlan about the article. The two began to brainstorm ways to bring Pleistocene rewilding some serious scientific attention. (It is worth noting that this conversation took place on a lunch break during a field study of mountain rattlesnake ecology, not a pursuit for the easily daunted.)

Greene and Donlan wanted to bring together a group of well-respected ecologists to discuss Pleistocene rewilding, with the goal of producing a scholarly article laying out its rationale and benefits. They began by submitting a proposal for such a meeting to the prestigious National Center for Ecological Analysis and Synthesis in Santa Barbara. The reaction was not encouraging. "They basically laughed us out of the room," says Greene. Still, he and Donlan persisted. Through a friend of a friend, they found a connection at the Ladder Ranch, one of several large tracts of rangeland owned by Ted Turner, the media mogul and conservationist. The manager at Ladder Ranch agreed to host the conference without charge. One weekend in September 2004, the invited guests converged on the ranch, in the desert outside of the town of Truth or Consequences, New Mexico.

The assembled researchers numbered an even dozen and included such committed rewilders as Martin, Burney, and Soulé; Joel Berger, a biologist with the Wildlife Conservation Society who had studied megafauna in Grand Teton, the Arctic, Africa, and Asia; and Jane and Carl Bock, grassland ecologists and old friends of Martin who were skeptical that the modern West truly held empty niches for exotic giants. Greene remembers the Ladder Ranch meeting as one of the high points of his professional life. "We had some very spirited arguments, but there was no ego jockeying, no bad feelings." By the close of the weekend, the group had produced a draft article on Pleistocene rewilding, which they intended to publish in either *Science* or *Nature*, the two top peer-reviewed scientific journals. Josh Donlan, the only student in the group and the meeting organizer, would be the lead author. Martin, by then well into his seventies, was happy to find a young ecologist to carry the banner for his cause.

When the article ran in *Nature*, it was accompanied by a graph that compared the conservation value and social conflicts involved in introducing African lions and cheetahs, elephants, camels, and horses. For Donlan and Greene, the lead authors, the fallout was intense. In the week after publication, they received about 1,000 emails, along with numerous requests for radio and television interviews. There was plenty of hate mail: one correspondent wrote succinctly, "If I see you or your elephant, I will shoot your ass." There were also many messages of support, including some from members of the National Academy of Sciences.

Greene, a Texas native, cherishes a note sent by an open-minded Texas rancher who wanted to know how he could help kick-start rewilding.

Josh Donlan, who has since completed his doctorate at Cornell and founded his own biological consulting firm, continues to write and speak in support of Pleistocene rewilding. He spends most of his time on islands scattered around the globe, helping to kill off invasive species. On islands the destructive force of the rats, pigs, goats, cats, and dogs people have introduced are painfully obvious. Many remote islands hosted a staggering diversity of birds and insects, but no mammals, in the millennia before humans arrived. In the Hawaiian archipelago, for instance, native species of plants and animals are now rare, swamped in an endless tide of invasives. Hawaii experienced two great waves of extinction: flightless ducks and rails vanished soon after the arrival of the original Polynesian settlers, who came onshore with rats, dogs, and pigs, all of them potent disruptors of habitat and predators on naive native birds. A thousand years later, Europeans arrived, bringing cats, mongoose, goats, cattle, and mosquitoes carrying avian pox and malaria. In 1893 the islands were home to sixty-eight species of native birds. Today, twenty-nine of those are extinct, and seventeen more are endangered.[4]

Donlan's specialty is designing eradication programs that kill every last rat, rabbit, goat, donkey, or pig on an invaded island. He has worked on islands in Australia, California, Mexico, and the Galapagos, using sophisticated computer mapping to help hunters find every last animal. These efforts can rescue island natives that had barely been hanging on. On Santiago Island in the Galapagos, the removal of many thousands of goats and pigs allowed the endangered Galapagos rail to rebound dramatically, the population growing from a handful to hundreds of birds over the course of eight years.[5]

It may seem strange that a man who spends much of his time battling introduced species would champion the idea of bringing Asian and African animals to the Americas. For Donlan, Pleistocene rewilding on a continental scale is a completely different proposition than introducing exotic mammals to islands. These invaders are utterly alien to island ecosystems and quickly devastate the native fauna. In contrast, America was a center of evolution for large mammals before and during the Pleistocene. Its landscapes, unlike those of remote islands, evolved with large, hoofed animals and their predators. Rewilders want to bring in close relatives of extinct Ice Age natives; in some cases, like that of the African lion, genetic evidence suggests it is the very same species that once stalked the Great Plains. The beasts they envision bringing to America's open spaces are relatively long-lived and slow to reproduce. "They're the opposite of rats and rabbits," says Donlan. "It's hard to imagine a runaway tortoise population, or a runaway lion population. What's more, we killed them once; if we have to, we can kill them again."

He points out that what Pleistocene rewilders have actually proposed is not as extreme as is sometimes portrayed by the press or perceived by the public. The idea is not to drive a van load of cheetahs and elephants to the outskirts of Topeka and let them loose on the plains. He envisions a long, carefully planned process, in which Old World megafauna would be introduced to spacious fenced enclosures and monitored, yielding information on their ecological impacts before any more drastic steps are taken.

Donlan, Soulé, and Greene are frustrated with the state of the conservation movement, which they feel has been reduced to exposing, and merely slowing, the loss of biodiversity. They consider the old idea that wild places can be returned to some perfect, static state to be deluded, condemning conservationists to endless defensive actions and to attempts to rescue species always teetering on the brink. It is time, they say, for a significant change in the way we value and understand wild nature.

A major lesson to come out of the last half-century of ecological science is that there is no single ideal condition for any landscape; with or without human influence, ecosystems are ever-changing. Yet, in a world where no remote corner has escaped the fallout of modern industrial chemicals or man-made shifts in climate, many conservationists still seek the restoration of an imagined state of pristine nature. In the Americas, that elusive goal is the landscape of 1492, before Columbus reached the New World. This vision relies on the discredited idea that ecosystems naturally reach and remain in stable states. It also ignores the impacts of Native American peoples who shaped the landscape by setting fires, hunting wildlife, and gathering and farming crops. Over many thousands of years, humanity has not only driven extinctions and triggered invasions of exotic plants and animals but also altered the course of evolution among many living species, a process impossible to reverse.

The necessity of awakening people to this reality had been on Soulé's mind long before he was invited to the Ladder Ranch conference. As early as 1990 he was predicting the birth of a new discipline of "recombinant ecology," which would explore the best ways to manage and conserve today's mixed-up ecosystems, where invasives, whether we like it or not, compete and coexist with native species.[6] He sees no way to undo the wave of human-driven species introductions. "What we can do," says Soulé, "is reverse the decapitation of the many ecosystems that have lost their most important players, the megafauna."

The Pleistocene rewilding proposal is deliberately provocative, and it sounds desperate, or plain nuts, to many. Some critics say that American ecosystems have changed so much over the past 10,000 years that it is now impossible for open spaces in the West to support any semblance of the Pleistocene megafauna. To others, the suggestion that we experiment with reshuffling big beasts among the continents—a process that the animals accomplished repeatedly in Ice Age times by crossing the Bering land bridge—is far too risky, given the track record of exotic species introductions. Old World animals might carry virulent new diseases to America or contract some western plague to which they have no immunity. There is no guarantee that elephants from the tropics could survive the cold of a Great Plains winter. Perhaps a more powerful point, raised by many involved in the uphill struggle to restore dwindling native species in the West, is that zealous proponents of Pleistocene rewilding seem to ignore some basic human realities.

People may be sparsely distributed on the Great Plains, where advocates of Pleistocene rewilding have suggested starting experiments with camels, cheetahs, and other Old World megafauna. Still, the small towns that dot the landscape rely on large swathes of land to run cattle and plant crops. Residents of the plains have encountered grandiose wildlife restoration plans before—in the 1980s, a movement to create a huge bison preserve covering 139,000 square miles stretching from Texas to Montana was quashed by local resistance. Yet restoration is sorely needed: the Great Plains today are empty of free-ranging bison, and black-tailed prairie dogs survive in only 5 percent of their historic range. These, like African elephants in Kenya, are both keystone species, animals that help create habitat needed by many other types of wildlife. The Nature Conservancy, the World Wildlife Fund (WWF), and regional conservation groups have been working for years to find common ground with local residents and fear that carefully built trust could be shattered by the very suggestion of rewilding African beasts onto American land. One staffer at WWF, which is working to restore native prairie species in north-central Montana, described the Pleistocene rewilding proposal as "the height of folly."

The restoration of wolves in Yellowstone, a shining example for the rewilding of native species, has been a remarkable success, in terms of both ecology and economics (tourists drawn to wolf-watch bring an estimated $35 million to the area each year[7]). But the top dogs remain a focus of controversy. As the population grows, some wolves roam beyond protected park habitat in search of new territory and wander onto ranchlands. When this happens, they sometimes prey on livestock. In 2008, wolves in Idaho, Montana, and Wyoming killed 184 cattle, 355 sheep, 14 dogs, and a handful of horses, llamas, and goats. The toll is low compared to the dire predictions made by some cattlemen's groups prior to the 1995 reintroduction, but it has not endeared wolves to ranchers. In response, federal animal control agents killed 264 wolves suspected of taking livestock in 2008, about 14 percent of the wolf population living in the U.S. northern Rocky Mountains.

Doug Smith, the lead wolf biologist in Yellowstone, has experienced firsthand the clash between wild predators and the locals. He lives with his family in Gardiner, Montana, a tiny town of 800 outside the park. Most of his neighbors opposed the return of the wolf—some remembered stories told by their parents or grandparents, who had had a hand in the killing of the last local wolves. Those ranchers of the 1920s and 1930s were proud of their accomplishment: at the time, wolves preyed heavily on livestock, because their natural staple, the elk, had been decimated by market hunting.

The 1995 reintroduction came only after years of intense political battle between pro- and anti-wolf lobbies. In the fifteen years that Smith has lived in Gardiner, wolf populations have thrived and rural people have come to a grudging acceptance of their presence. Still, everywhere he goes—the grocery, the high school football game—people want to talk wolves, and he is often called on to defend the predator's right to exist. "Ranchers and hunters would rather have things back as they were fifteen years ago, before we brought the wolves back," says Smith. "But most realize the only way to go forward now is to make some allowances.

It took trench warfare to get the wolves in here, and it took ten years for things to calm down; now we're out of the trenches, talking." He worries that the idea of Pleistocene rewilding will send local people back to their political bunkers. There is no sense of compromise in a vision of wild elephants and lions as the next big step in American conservation. "What you get is a loss of good will, a loss of political capital. There are more important issues to spend that capital on: restoration of native species, protection of wildlife habitat corridors."

Donlan sees the clash between people and large predators as a question of environmental justice. "Do we expect Africa and parts of Asia to bear the burden of conserving the planet's megafauna by themselves?" he asks. "Americans seem to have a 'no lions in my backyard' attitude." Wyoming ranchers no doubt feel fully entitled to that attitude—they're already disgruntled about having to cope with wolves, which sometimes attack livestock, but not people.

Yellowstone has often been described as the American Serengeti. On the edges of the real Serengeti National Park, in Tanzania and Kenya, however, large predators affect poor farmers in ways that make the wolf–rancher clash in the northern Rockies look trivial. One study in Tanzania found that livestock depredation accounted for the loss of 20 to as much as 60 percent of annual income among already impoverished farmers on the outskirts of Serengeti National Park.[8] The culprits were nearly always spotted hyenas, the most numerous predator in the area and the one best adapted to commute from protected park habitat into human communities. But retaliatory killings by angry farmers target other carnivores too, including endangered leopards and lions.

As human populations climb in Africa, lion numbers dwindle—the world total has plummeted to 50,000 animals in recent years. Outside of national parks and game reserves, the big cats have been eradicated from most of their range. One exception was a stretch of coastal scrubland in southern Tanzania, near the Selous Game Reserve. In a study that by ironic chance ran in the same issue of *Nature* as the Pleistocene rewilding proposal, a group of researchers led by Craig Packer of the University of Minnesota documented patterns of lion attacks in the region.

More than 563 people were killed and 308 injured in lion attacks from 1990 to 2005. The number of attacks increased over the course of those years, a result of a rise in the number of people living and farming in lion habitat.[9] Normal prey species such as wildebeest and zebra do not survive the conversion of wilderness to cropland. So lions instead go after bush pigs, a native species that thrives in Tanzania's farm fields. Farmers often sleep in makeshift huts built among their crops, to guard against the night raids of hungry bush pigs. Lions also forage by night, and on occasion, instead of capturing a pig for dinner, a cat will drag a man out of his hut.

Farmers in southern Tanzania are among the poorest people on earth. No one owns a flashlight, let alone a rifle. Packer and his colleagues documented chilling tales of children snatched from their mothers' arms and women pulled from their beds at night. Still, people have discovered an affordable, effective way to fight the lions. A few years ago, an elderly woman rose from her bed to go to the outhouse. When she did not return, her husband went to look for her and found the upper half of her body lying on the ground. He fetched a box of rat poison, which he used to lace his wife's remains. The lion that returned to feed on the woman's body died, and the same grim strategy worked a year later for another man who found, and poisoned, the lower half of his mother-in-law's body. In 2009, Packer reported that most of the lions had disappeared from rural southern Tanzania, at least for the moment.[10] It is only a matter of time before new lions spread out of the wilds of the Selous Reserve. Packer continues to work on ways to help local people keep bush pigs out of their fields. If such efforts succeed, they will protect crops and farmers as well. Without bush pigs to draw them in, lions will stay farther from the places where people live and work.

Most Americans are unaware of these incidents. Still, setting African lions loose in the western United States is a nonstarter, a reality that even the most zealous proponents of rewilding understand. "We have to be politically as well as ecologically realistic," says Soulé. "We might get some camels back in North America, but we're not going to have lions roaming around because of the conflicts that would provoke with human beings." Greene, who gauges public reaction at his lectures on rewilding, agrees that the African lion is a deal breaker. He still has visions of importing cheetahs, which are much lighter than lions, go after smaller prey, and do not attack people. African cheetah populations have shrunk dramatically over the past forty years, and the largest group of survivors—about 2500 animals—live on agricultural lands in Namibia.[11] There, conservationists encourage farmers to use guard dogs to protect their flocks rather than preemptively killing passing predators—a strategy also being used by some ranchers in the United States to deter wolves from preying on their stock. Greene hopes to see cheetahs transported from their shrinking African range to pursue American pronghorn in the wild. Recent genetic studies of fossil American cheetahs show that they were not closely related to their African counterparts.[12] Still, the two kinds of cheetahs were built in much the same way. Alan Cooper, an expert on ancient DNA who worked on the American cheetah study, agrees that the African form might make a good living hunting pronghorn and mule deer rather than springbok and impala. He doubts he'll ever witness the experiment, however. Modern cheetahs might fit seamlessly into the ecosystems of the West, but the resurrection of such a long-extinct predator is unlikely to be accepted by local people.

Fig. 22 Some advocates of Pleistocene rewilding dream of bringing African cheetahs to the western United States, to replace the extinct American cheetah, which coevolved with the speedy pronghorn. (Photo by James Temple, from Wikimedia Commons, http://en.wikipedia.org/wiki/File:TheCheethcat.jpg)

Even native wildlife can cause alarming ripples of change when they move into new territory. In recent decades, white-tailed deer, which once roamed only east of the Mississippi, have invaded the West. Irrigated fields of alfalfa and sweet clover in the Columbia River basin have created ideal whitetail habitat where it never existed before. During the winter, white-tailed deer stay down in the valleys and don't have much impact on mule deer, which are native to the western mountains. But in summer, the whitetails migrate out of the low country and into higher elevations. Cougars follow white-tailed deer into places they would otherwise seldom visit. There they prey indiscriminately on both species of deer. In the Selkirk Mountains, where the boundaries of Idaho, Washington, and British Columbia meet, the resulting influx of cougars has caused mule deer to dwindle and threatens a highly endangered population of mountain caribou, the last in the United States.

To protect mule deer and caribou, state wildlife managers decided to intensify cougar hunts in the Selkirks. The strategy backfired, however, because it failed to account for the subtleties of cougar society. Hunters target mature male cougars, the most prized trophies for humans. Without adult males to keep the younger generation in line, young males kill infants they did not father, a strategy that makes females available sooner for mating. Given this risk, a mother cougar would rather move than lose her young. Robert Wielgus of Washington State University's Large Carnivore Conservation Laboratory found that heavy cougar hunting thus led to a greater loss of the species the humans had intended to protect. Female cougars, pushed by an influx of infanticidal males, moved to higher elevation. There, the mother cats preyed heavily on mule deer and caribou.[13]

Any sort of rewilding, whether it focuses on the time of Columbus or the era of the mammoth, will ultimately hinge on finding ways for humans to coexist with big predators. So far there has been little success. Most wildlife managers still believe the long-standing dogma that hunting adult males will not affect the viability of a predator population, though Wielgus has documented a syndrome of destabilized social systems, leading to a decline in reproduction, among grizzlies as well as cougars.[14] Another dogma holds that wolves, which reproduce much faster than most large carnivores, can survive intense hunting pressure. While this is true, studies of wolves in protected areas, including new

Fig. 23   The cougar, like the wolf, boosts biodiversity by controlling populations of its prey. By disrupting the predator's social structure, intense cougar hunting in Washington and Idaho has led to an increase in cougar conflicts with people and diminished numbers of threatened mountain caribou and mule deer. (Photo by Tim Knight)

work on the Yellowstone wolves—where some individuals live to the ripe old age of ten years or more—make it clear that wolf societies can be shattered by heavy harvest.[15] Wolves living in stable packs learn sophisticated hunting techniques from their elders, enabling them to hunt more efficiently. By contrast, in places where hunting is permitted, wolves seldom survive more than three or four years and tend to live in small, shifting coalitions instead of cohesive family groups.

There are good reasons to keep wolf social structures intact: small, unstable packs kill more large prey animals per wolf, and they are less likely to boost biodiversity than are the large, family-based packs in Yellowstone's interior. In October 2009, a few weeks after Montana opened its first legal wolf-hunting season in decades, a hunter a mile outside the boundary of Yellowstone National Park shot and killed the alpha female of the Cottonwood Pack—the leader of a wolf family that had lived almost invisibly in a remote corner of the park. In the aftermath of the female's death, her pack fragmented, and several other members were also shot. A representative of Montana's Department of Fish, Wildlife, and Parks claimed the hunt had been designed to target wolves that preyed on livestock—but the Cottonwood wolves had never gone near a ranch.[16] Ripple and other wolf researchers now advocate a no-hunting buffer zone around the park to protect wolf packs that spend most of their time inside Yellowstone.

Scientists are only beginning to understand the ways we affect the cougar and the wolf when, as often happens, we manage them based on politics and fear. The naked apes of North America are not ready for African lions—but no other living four-footed predator could take on an elephant, a camel, or a wild horse. Even Donlan, who writes movingly of bringing the Pleistocene megafauna back to life, knows that his vision rests on the backs of big carnivores. To understand why, you need only look at America's most prominent example of unplanned rewilding—the passionately loved, hotly resented wild mustangs of the Great Basin.

WILD
REALITIES

RANDOM ACTS OF REWILDING

A black stallion, his coat gleaming in the sun, stands guard over his harem of nine mares. The mustangs are graceful, small-footed, and compact, built to survive in rough terrain. Their brown and gray coats contrast with the deep blue of sky and the bright spring blossoms of redbud. New mothers walk slowly among scattered chunks of volcanic rock, their young foals trotting at their sides. One of the babies, says Dianne Nelson, the owner and manager of the Wild Horse Sanctuary, was born this very morning.

More than thirty years ago, Nelson made an impulsive decision that would change the course of her life. There had been a mustang roundup on U.S. Forest Service land near her home in the Sierras, and the agency planned to kill eighty horses that no one wanted to adopt. Nelson decided to take the animals rather than see them destroyed. She had nowhere to put them, but other wild horse advocates rallied to the cause, donating the use of ranchland and offering sponsorship money. Today Nelson's nonprofit, Wild Horse Sanctuary, shelters about 300 mustangs on 5,000 acres in the shadow of Mt. Lassen, in Northern California. A smart, pragmatic ranch woman, Nelson finds the idea of Pleistocene rewilding incomprehensible and does not seem to grasp the connection to her sanctuary work. Yet if there ever existed an unplanned test of rewilding, it is the case of the mustang.

The last native American horses died out at the end of the Ice Age, along with the mammoth and the saber-toothed cat. Long before, horses had migrated west over the Bering land bridge and given rise to an array of Old World cousins, including the European tarpan, the Przewalski's horse of Asia, and the African zebra. Sometime after the first Paleoindian people reached the Americas, humans in Europe and Asia began to tame the horse. Columbus brought domesticated horses to the island of Santa Domingo in 1493, on his second voyage to the New World. Nearly every ship that left Spain for America in the early years of sixteenth-century exploration carried horses—both the stallions the conquistadors preferred to ride and mares to populate the new land.[1]

Explorers such as de Soto and Coronado brought hundreds of horses with them on expeditions to the places we now know as Florida, the Midwest, and the southwestern United States. Most of these animals became food for starving expedition members, and the few that were released were quickly butchered by Indians. It wasn't until 1598, when the Spanish established a permanent settlement in the upper Rio Grande Valley of modern-day New Mexico, that North American Indians began to ride horses rather than eat them. The horse culture of the Plains Indians was born—and as horses escaped to form feral bands, so was the mustang.

Most mustangs today live on federal lands in the Great Basin, the vast arid landscape between the California Sierras and Utah's Wasatch Range. They carry genes from old Spanish horses and Indian ponies and from work horses that belonged to pioneers heading west in the nineteenth century. Those that have survived in the wild are hardy and intelligent, strong of foot and leg, and free of the digestive and birthing problems that plague purebreds. Nelson uses tamed mustangs as saddle horses because, she explains, they are the healthiest and easiest to keep.

As domestic cattle and sheep came to dominate the western range, they squeezed out mustangs, along with native species like bison, bighorn sheep, and pronghorn. Today more than half of the wild horses in the West live on Bureau of Land Management (BLM) property in Nevada. This is a historical irony: the BLM began its existence as the Grazing Service, under the Taylor Grazing Act of 1934, a piece of legislation meant to favor cattle ranchers. In its early decades, the BLM did much to kill off mustangs, then as now seen as unacceptable competition for forage by ranchers who run livestock on public land.[2] Yet under the Wild Free-Roaming Horse and Burro Act of 1971, a political counterstrike pushed through by wild horse lovers, BLM is now legally bound to protect and benignly manage mustangs.

Advocates like Nelson don't want to see wild horses shot to make way for cattle and sheep. Domestic livestock that graze on public lands number in the millions, she points out, and have far more impact on the landscape than do the mustangs, which number in the tens of thousands. All true, yet the legal and political status of the wild horse remains disconnected from its ecological reality. Mustangs are not managed (and routinely slaughtered) like cattle, and unlike mule deer, elk, pronghorn, and bighorn sheep, they cannot be legally hunted. Thus, horse numbers boom, overburdening the available habitat. Native wildlife suffers as a result, and so do the mustangs themselves—in times of drought or overcrowding, they slowly starve.

Erik Beever, a biologist with the U.S. Geological Survey, has studied the impacts of mustang herds at sites scattered across the Great Basin. Horses trample and compact soils, destroy streamside vegetation, and diminish the numbers and diversity of native plants, mound-building ants, reptiles, and small mammals.[3] They spread invasive grasses whose seeds cling to their rough coats or pass through their guts. Domestic livestock are also guilty of these sins, but horses remain on the range year-round, unlike cattle, which are moved with the seasons.

Beever's work suggests that mustangs may be less of a threat to cattle than most ranchers fear. Horses prefer high ridges and spend most of their time at elevations cows seldom reach. They inhabit the same mountain spaces as pronghorn and mule deer, but these creatures are shrub eaters while the horse is a grazer. However, mustangs can reshape an ecosystem and may affect native herbivores in subtle ways.

Wild horse herds expand at a rate of 20 percent per year, which explains why BLM counts of mustangs ballooned from 17,300 in 1971 to more than 50,000 in 1976, a mere five years after the Wild Horse and Burro Protection Act was passed. The lack of predators to control horse populations is a problem: the massive Pleistocene carnivores—the saber-toothed cat, the American lion, and the dire wolf, all of which likely preyed on horses—are long gone. Coyotes and mountain lions, the top predators over most of the mustang's range, do not make a real dent in wild horse numbers. Now and then a cougar will hunt a foal, but adult horses are essentially immune to nonhuman predators.

At the Wild Horse Sanctuary, Nelson has preserved what wildlife habitat she can, given her mission of mustang rescue. The redbud and

Fig. 24 Wild mustangs rounded up by BLM workers, Susanville, California, 1984. (Photo by Katey Barrett, copyright kateybarrett.com.)

buckeye stand tall, and abundant woodpeckers feed their chicks, who wait safe in the crevices of old oaks. But every blade of grass has been cropped to invisibility by hungry horses. Most seasons of the year Nelson must feed them hay, an expensive proposition. She is hustling to raise money to build fences on the ranch, making it possible to rotate grazing pressure and let grass rest and regrow.

"We do need to manage horses, because we've taken all the natural predation out of the mix," says Nelson. In tones of regret, she tells of a pretty young filly, killed and eaten by a cougar just before she was scheduled to leave for her adoptive home on a California ranch. Cougars are part of the natural order, she acknowledges wryly, until they come into her backyard. Yet, like other wild horse advocates, she is unwilling to allow humans to step in and play the role of top predator. Instead, she thinks BLM should put up for adoption as many young horses as possible and treat wild mares with PZP, a contraceptive vaccine that prevents pregnancy for two years or more.

BLM rounds up mustangs when populations grow beyond prescribed numbers in government-designated horse range. Some are adopted, and the rest end up in holding pens—a miserable situation for the horses and a major expense for BLM. Nelson and others vigorously protested a plan to euthanize excess horses, and BLM declared a temporary moratorium. The horse's powerful mystique has given it a status enjoyed by no other large animal in the United States, native or introduced. Decades ago, it was common for mustangs to be rounded up and sold to slaughterhouses for pet food; under a federal law passed in 2007, it is now illegal to slaughter horses and process their meat for consumption.

Mustang advocates find Pleistocene rewilding—with its introduction of large predators—untenable. Harry Greene feels the same way about what he sees as a crusade to protect wild horses at any cost. The rewilding paper in *Nature* specifically discussed the integration of horses into a resurrected Ice Age ecosystem, adding that in genetic terms, the horse that now roams the West is essentially the same species that disappeared from America 13,000 years ago. After the article ran, e-mails began pouring in from wild horse advocates, asking Greene for more information to back up their view that the mustang is a natural inhabitant of the West. As soon as Greene mentioned that he envisions horses becoming prey for big carnivores, his correspondents vanished into cyberspace.

"I don't know what they think will happen to these horses," muses Greene. "If they're not rounded up, controlled and slaughtered by the BLM, they won't ascend painlessly to heaven. The realistic possibilities are that they die of disease and starvation, or something kills them and eats them. We live more and more in an enchanted illusion of what nature is, which I think is counterproductive to conservation." At his talks on rewilding, Greene makes a point of showing slides of wolves and lions feeding, their faces smeared with blood.

In the 1990s, researchers came across a rare example of a horse herd whose numbers were holding steady. In the Montgomery Pass Wild Horse Territory, on the California–Nevada border, mountain lions keep the mustang population in check. They seldom tackle adult horses, but the cats prey efficiently on foals. In a typical year the cougars kill about half of the foals born. In most other wild horse habitats, 70–97 percent of foals survive their first year.[4]

John Turner of the University of Toledo in Ohio has been tracking the Montgomery Pass herd, and the cougars that prey on them, for more than twenty years. What stabilizes the balance there is the presence of a migratory herd of mule deer, he explains. The deer migrate west out of the pass into the Sierra high country in the spring. As they leave, mustang mares begin to birth their foals. A contingent of cougars at Montgomery Pass have honed a strategy in which they switch from hunting deer to hunting newborn horses in the spring. By autumn, the year's foals are too big for a cougar to tackle, at which point the mule deer return. Mother cougars teach this carefully timed foraging behavior to their young.

Montgomery Pass is the only horse territory in the West where there is no need for population control, and thus the only place where mustangs can lead truly wild lives. A single stallion may dominate a band for many years, and young horses grow to adulthood in their natal groups. That rarely happens in other areas, where humans round up excess horses every couple of years, disrupting existing families. This undisturbed state allows horses to develop traditions and pass them down. Some bands, for example, have chosen to stay at low elevations, where cougars are scarce, to reduce the risk of predation. Others continue to use the high country, coexisting with the big cats.

The ecological balance at Montgomery Pass is unlikely to be replicated elsewhere in the Great Basin; in less remote places people will not tolerate the cougar density needed to keep horse populations in check. Without effective predators, though, wild horses come to a bad end. Many starve after their herds devour all the grass; habitat for native grazers is damaged in the process. Still, judged by its ability to adapt, the mustang is an outrageously successful example of American wildlife. It thrives in rugged, dry country where other creatures could not survive. Flying by helicopter in the course of his work, Turner has hovered close to horses moving at full gallop over rough ground. "I used to worry that they'd hurt themselves, but the darn things just don't. They'll run across unbelievably rough, rocky, steep terrain, and just go. They're so well integrated with that environment... that's why they're such amazing creatures."

The unplanned rewilding of horses in America has become an ecological dead end, due to human passions and politics. The threat of a hungry carnivore lurking at the water hole is the essence of the truly wild horse. In the absence of large predators, mustangs diminish, rather than enhance, biodiversity.

Another candidate for Pleistocene rewilding is the camel, which has been on Paul Martin's list since the 1960s. The two species of living camel, the double-humped Bactrian and the single-humped dromedary, are descendants of American ancestors who crossed into Asia millions of years ago. Domesticated dromedaries are common in Somalia, the Sahel, the Middle East, and southern Asia, but none survive in the wild. A thousand or so wild Bactrians hang on in the Gobi Desert of China and Mongolia; many more live as beasts of burden.

The camel has no place in the iconography of the American West. That might make it a better choice than the horse for a formal rewilding experiment—no one idealizes the beast as many do the mustang. And modern camels have already shown they can thrive in the deserts of the West. The Spanish conquistadors returned the horse to its home continent; Jefferson Davis, while serving as Secretary of War under President Pierce in the 1850s, pushed to bring camels to the West. Camel caravans under the command of Colonel Robert E. Lee, the future Confederate general, brought feed and water to supply horses kept at remote Texas outposts. Unlike horses and mules, the camels could live off the desert without supplemental rations.[5]

In 1856, twenty-five dromedaries carried supplies for a military survey that charted a new road from Fort Defiance, New Mexico, to the Colorado River in California. The road still exists as a section of Route 66. Edward F. Beale, who led that first American camel caravan, described the animals as docile and easily fed. "They appear to prefer to browse on the mesquite bushes and the leaves of a thorny shrub, which grows in this country everywhere, to the finest grass," he wrote.[6] Beale thought the camels a great success, but the Army abandoned its Camel Corps with the onset of the Civil War.

Americans were not accustomed to camels, with their strong smells and piercing wails of complaint. A popular book on Asian travel, published in the 1840s, described a camel as having two defenses against potential predators: "Its huge shapeless ugly frame, which resembles, at a distance, a heap of ruins... and a vehement sneeze, wherewith it discharges, from nose and mouth, a mass of filth against the object which it seeks to intimidate or annoy."[7]

Aesthetics aside, the dromedaries proved to be mighty pack animals, able to bear twice as much weight as a typical mule while traveling farther and faster. The military camels were eventually auctioned off and used by civilians to pack supplies to remote mine sites in California and Nevada. A German-American entrepreneur named Otto Esche imported Bactrian camels for this very purpose; some of the animals he shipped to America worked as far north as British Columbia. The camels could live well on creosote and other desert shrubs that horses and cattle would not touch. But they were despised by teamsters who managed mule trains. The sight and smell of a camel spooked horses and mules, sending them into uncontrollable, wagon-crushing panic. In 1875, Nevada outlawed free-roaming camels on its public roads. The cost of coping with incensed horse owners came to outweigh the camels' strength and resilience as work animals. At the close of the 1870s, nearly all the imported camels were turned loose to fend for themselves. By the early twentieth century the last of them had died out.

In a letter to *Nature,* one critic of Pleistocene rewilding claimed that the fade-out of introduced camels was proof that western ecosystems have changed too much in the past 10,000 years to sustain Ice Age megafauna. Habitats have indeed evolved, but this argument ignores the reality of the relationships among people, camels, and horses. Small herds of camels survived for decades in western deserts. Feral camels were once numerous along the Gila River in Arizona, but due to hunting they had vanished by the 1890s. Camels were reported to be roaming near Goldfield, Nevada, in 1905. As late as 1940, a man living in the Arizona ghost town of Quartzsite, along the original route of Beale's expedition, claimed to have found wild camel sign in the desert nearby. He told a bemused Associated Press reporter that he planned to bring home camel meat for Christmas dinner.[8]

The most common response to the sight of a wild camel in the West was to shoot at it. It took horses many decades, during which they were nurtured by humans, before they established resilient populations in the wild. Fewer than 400 Old World camels were brought to the American West and later released in small, scattered groups that were never given a reasonable chance to endure. But the camel is an adaptable beast, and abundant proof of that exists in the deserts of Australia.

A few nineteenth-century Americans dabbled in the importation of camels, but their contemporaries in Australia took up the idea in a big way. The first camels arrived in Australia in 1840; by 1907 up to 20,000 dromedaries had been brought to the continent's arid interior, for use as mounts and as pack and draft animals. With the coming of automobiles in the 1920s and 1930s, camels lost their value. Some were slaughtered for food, but a great number of them—as many as 10,000—were turned loose in the outback.[9] Able to go for days between drinks of water, feeding on native desert plants, they thrived.

Moreover, the camels appear to fill a unique ecological slot. During the two decades they have devoted to the study of Australia's feral camels, biologists Birgit Dorges and Jurgen Heucke found that native shrubs seem to have adapted to the effects of large animal browsers—an impact that had been missing since the extinction of the native megafauna. Acacias and whitewood fed on by camels produce more leaves and appear healthier than those left untouched. Dorges and Heucke suggest that the camel fills a niche left empty by the passing of *Diprotodon* and the giant short-faced kangaroos.[10] The acacias have long memories—in large enclosures set up to study camel ecology over a period of years, acacia bushes developed defensive spikes up to four-tenths of an inch long. Where camels are newly arrived, acacia thorns are much smaller or absent, and in places that lack both camels and cattle, the only large browsers in modern Australia, the acacia bush grows no thorns at all. In Queensland, camels are used to control the growth of invasive shrubs introduced from Africa and America. Hungry camels accomplish what pesticides and digging cannot, clearing out even prickly weeds that cows and horses will not touch.

Australia's camels also offer a unique glimpse of the lifeway of wild dromedaries, which exist nowhere else on the planet. Highly social giants—Dorges comments that except for elderly bulls, camels rarely spend time alone—they live in groups where mothers share the responsibility of minding their young.[11] During the breeding season, dominant males take command of these groups and fight competing bulls for the right to mate. Pregnant mothers ready to deliver their calves will seek seclusion, because bulls often kill newborns. Among camels, as among cougars, males practice infanticide as a strategy to bring females back into estrus quickly.

A species returned to the wild and acting as a partial ecological replacement for extinct native giants: it might appear that the camel in Australia is an example of successful, if accidental, Pleistocene rewilding. But there is a hitch: Australia's stock of large carnivores is even scantier than North America's. The marsupial lion and the giant lizard are long gone, and no living predator, short of humans with guns, can tackle the camel. Their unimpeded growth is formidable—the population doubles every nine years. In the early 1980s, when the first formal survey was conducted, about 43,000 camels lived in the outback. By 2008 that estimate had shot up to one million. Camel herds now roam across most of the country's interior, in densities that can bring the growth of native trees and shrubs to a halt. Though camels seem able to survive on almost any growing plant, their booming numbers threaten the native shrubs that they find most delectable, including the quandong, bean tree, bush banana, and bush plum. These plants produce fruits that are prized by Aboriginal people.

In January 2007, after several years of drought, central Australia reached a peak of dryness. Starving, thirsty camels began to move into areas of human settlement in search of water. Tens of thousands of the animals invaded Aboriginal settlements and pastoral lands in West Australia, South Australia, and the Northern Territory. The small Aboriginal community of Docker River in the Northern Territory had more than 500 camels roaming its dusty streets for weeks. Camels broke open water taps, shattered toilet bowls, and wallowed in the water that leaked out. They died of hunger in upstream water holes, polluting the water sources for small outback towns.

The response among white Australians was to kill the herds off. Thousands of wild camels were shot in early 2007. Some of the bodies were processed into pet food, but about 4,500 were shot and left to rot. Aboriginal people in Docker River resisted an expensive plan to hunt camels by air, but months into the camel siege they agreed to a cull of the area within 50 km (31 miles) of their town. That camel hunt never happened. The day before the first helicopter flight was scheduled, rain fell over much of the interior, and the camels melted back into the remote deserts. But two years later the town experienced an even more intense camel invasion. Many hundreds of the animals smashed water supply pipes and milled around the resulting water holes. In December 2009, a government-hired marksman shot about 3,000 camels in and around Docker River.[12]

Everyone who studies camels Down Under agrees the population must be managed by humans. Glenn Edwards, a biologist with the Australian National Parks and Wildlife Service, says that the camel problem will only intensify unless Australians act to change the situation.[13] Uncontrolled camel populations could drive rare desert plants into extinction, and native wildlife may follow.

Dorges and Heucke propose that camels be raised alongside cattle, increasing the productivity of Australia's arid rangelands. There is a growing market for camel meat and milk in Asia and Africa, and camel steak is tasty and lower in cholesterol than beef. Ideally, the researchers would like to see the camel replace the cow—it is far better suited to desert life. A fledgling industry exports live animals to Malaysia and Brunei, where they are used for racing, and markets camel meat, which is slowly being accepted by Australians. Perhaps the camel problem can someday be transformed into an asset, a sustainable means of income for impoverished Aboriginal communities and struggling outback ranchers. For now, the costs of repairing the damage that feral camels cause to fences and water sources, as well as controlling camel numbers with aerial hunting, far outweigh the profits from selling camels, live or dead.

Booming camel populations add to the toll taken by introduced species in Australia, from the rabbit and red fox to the sheep and domestic cow. The only surviving native carnivores are small, and many are endangered: the polka-dotted quolls, the phascogale, which resembles a miniature opossum with an ostrich plume tail, and the numbat, which specializes in catching termites with rapid-fire movements of its tongue. Petite relatives of the kangaroos, including bettongs and hare-wallabies, have also been hard-hit. A growing contingent of ecologists now see the dingo, a predator introduced to the continent thousands of years before the first Europeans arrived, as the best hope for preserving dwindling native species. The dingo's most important trait, say supporters, is not its introduced status but its skill as a hunter of other exotic creatures.

Adam O'Neill, a muscular man with close-cropped hair, is a hired gun in the cause of Australian conservation. His battered Toyota pickup, fitted with a rifle rack, has seen thousands of miles of use in the remote outback. Landowners who want to protect or reintroduce native marsupials on their property call on him to clear out populations of invasive rabbit, cat, and fox. After years of this work, O'Neill has become a passionate advocate for the dingo. "The native mammals that still survive," he says, "do so under the protective influence of dingoes."

Australian society has been at war with the dingo since the 1830s, when the first settlers began to raise livestock in the forbidding, arid interior. Dingoes found sheep and calves easy prey, and settlers quickly concluded that the only good dingo was a dead one. At the turn of the twentieth century, construction began on one of the most ambitious projects of the predator-averse era, a cross-continental barrier designed to keep dingoes out of sheep pastures in southern Australia. Today, the "dog fence" stretches more than 3,100 miles across the desert—the longest fence on the planet. On either side of this man-made divide, ranchers, government officials, and even conservationists routinely shoot and poison dingoes.

O'Neill's interest in dingoes began when he was working far to the south of the dog fence, in an area where they had long been thought extinct. To his surprise, not only were dingoes living there, but they appeared to be protecting rare native marsupials. O'Neill had been hired to eradicate invasive mammals from a series of properties in South Australia, fenced off and destined to become sanctuaries for indigenous wildlife. To start, O'Neill fed the foxes. Free food lured them onto the refuge. Soon the area held a high density of foxes, which proceeded to kill off rabbits and scare any cats away. The foxes eventually became so tame that they would eat cheese from O'Neill's hand. Then, he left for two weeks. When he returned, every last fox had mysteriously vanished.

At first, O'Neill assumed the landowner had set out baits to poison the foxes—a routine tactic among many Australian conservationists, albeit one that he considers a serious mistake. It took some time, and repeated occurrences at other sites south of the dog fence, before he understood what was happening. Foxes would vanish when dingo signs began to show up, in landscapes that had been thought to be dingo-free for decades. "Dingoes were coming in, pissing on the feed stations I'd set up for foxes," he says. "They'd never eat it themselves; they just don't like so many foxes around, getting free food." South of the fence, dingoes have learned to live in ways that make them all but invisible to humans, but a bit of scraped-up earth spattered with urine spoke loud and clear to the foxes. Like gray wolves that attack coyotes on their turf, the dingo, Australia's top dog, shows little tolerance for its smaller cousin.

O'Neill thinks that dingoes south of the fence have developed a sophisticated culture of survival by stealth, passing it down from one generation to the next. In a landscape rich with sheep, dingoes seldom hunt livestock, though historical records show they are capable of doing so to devastating effect. Instead, they survive on a diet of rats, rabbits, birds, and snails. In places where intense poisoning campaigns are set up, dingo predation on livestock can actually increase. This may happen because poison baits kill off older, wiser dingoes, leaving inexperienced youngsters to fend for themselves. Similar impacts of predator control programs have been documented among wolf packs in North America.[14]

Fig. 25   An adult dingo and pup relaxing in the Australian outback. Introduced 4,000 years ago, dingoes now play the vital role of top predator in the continent's ecosystems. (Photo courtesy of Arian Wallach and Adam O'Neill.)

In Australian wilderness areas where dingoes are unmolested, they live in stable packs like those formed by Asian dingoes or North American wolves[15]. Little is known about dingo societies near towns or cattle stations, but dingoes in these areas howl less often and keep out of sight.

O'Neill works with Arian Wallach, a doctoral student at the University of Adelaide, collecting data to support his ideas about dingoes and their role in Australia's ecology. The two researchers set up transects in habitats that shelter rare remnant populations of yellow-footed rock-wallaby and carefully searched out tracks and spoor at sites on both sides of the dog fence. Once a common sight, the wallaby has been devastated by fox predation and competition for forage from introduced goats. O'Neill and Wallach found that dingoes coexist with wallabies on both sides of the fence—where dingoes walk, foxes are rare, supporting the theory that the dingoes offer protection from fox predation. In addition, most wallaby colonies are found within 1.25 miles of a water source. Dingoes need to drink up to twice daily, so they are tied to water sources; foxes, by contrast, can survive without free water. Nobody knows for sure how often yellow-footed rock-wallabies need to drink, but they are well adapted to arid Australia. They may be sticking close to water sources to stay near the protective influence of dingoes.[16]

Dingoes also appear to protect the malleefowl, an eccentric bird that builds large mounds of leaf litter, using heat generated by the composting process to incubate its eggs. Malleefowl survive in only a few patches of habitat far south of the dog fence, yet every site where O'Neill and Wallach found active nests held evidence of dingoes as well: their scats, tracks, and places where they had scent-marked malleefowl nest mounds.

Dingoes first reached Australia about 4,000 years ago, arriving with Southeast Asian seafarers. These animals were a barely domesticated variety of the Indian wolf. Unlike domestic dogs in western Asia, which by then had been intensively bred for a variety of unwolflike characteristics, dingoes looked, and often acted, like their wild ancestors.[17] Aboriginal people adapted to the new arrivals, keeping them as pets, helpers in kangaroo hunts, and, in a pinch, a supplemental food source. Meanwhile, dingoes quickly went feral and began to compete with Aboriginal hunters for kangaroos and other prey.

When the dingo landed, long after the mass extinction of Australia's megafauna, the continent held one last large marsupial predator, the thylacine. A wolfish-looking beast with striped hindquarters, the thylacine was also known as the marsupial tiger. It vanished from the mainland soon after the dingo was introduced but survived into the twentieth century on the island of Tasmania. Historical records describe thylacine behavior and document the intense bounty hunting that destroyed the last populations. Students of ecological history have long assumed that the dingo drove the thylacine to extinction on mainland Australia. The linchpin of this argument is a bit of circumstantial evidence: the thylacine disappeared at about the time the dingo arrived but survived in Tasmania's dingo-free landscape. For ecologist Chris Johnson, who has made a specialty of studying Australia's mammal extinctions, this picture is too simple.

Twentieth-century accounts of thylacines portray them as solitary creatures that hunted alone. Yet these are records of the last of their kind, members of an intensely persecuted population. Older reports, from early in the days of Tasmania's settlement, describe thylacines hunting in groups, "with the pertinacity of a pack of wolves on the steppes of frozen Russia."[18] They were fast, nimble, and big—more massive than a dingo and strong enough that, in at least one case, seven domestic hunting dogs could not kill a lone individual. A recent analysis of the bite force thylacines exerted, based on studies of their skull and jaw bones, suggests that thylacines took prey bigger than themselves.[19] All this conflicts with the idea that the dingo pushed the thylacine into oblivion.

Johnson suggests that intensified human hunting severely reduced thylacine populations in northern Australia before the dingo ever became established, pointing out that while both dingoes and thylacines were often depicted in rock art in the north, the Aboriginal people who first painted dingoes worked in a distinct style, more recent than any known image of the thylacine. The dingo may have played a minor role in the thylacine's disappearance; perhaps dingoes and humans hunting together became a truly formidable enemy. Regardless of how exactly the thylacine died out, says Johnson, the beast is gone, and the dingo is the only large predator left on the continent. He argues that native wildlife, and perhaps even outback cattle ranchers, are far better off with dingoes than without them. (In the absence of dingoes, kangaroo populations can skyrocket, leading to overgrazing of pastureland.)

In an analysis of threatened and extinct marsupial species across the continent, Johnson found that small mammals survived much longer in areas with healthy dingo populations. Where dingoes have been removed, 90 percent of ground-dwelling mammal species have gone extinct, due to predation by foxes and cats.[20] Johnson's interest in the protective effect of dingoes began years ago, when he was studying wallabies on an outback cattle station in northeastern New South Wales. The place was a haven for bettongs and rat kangaroos, threatened small marsupials. The landowners believed the bettongs thrived because they chose not to poison local dingoes, and the dingoes kept the foxes away. Soon after this experience, one of Johnson's colleagues was in the Tanami Desert, researching a threatened species, the rufous hare-wallaby. Dingoes in the area were occasionally eating hare-wallabies, so the local parks and wildlife service decided to poison them. "Within two weeks," says Johnson, "there were foxes on the site, and they killed off the hare-wallabies. That species is now extinct on the mainland."

The sad fate of the rufous hare-wallaby is a classic example of mesopredator release. "In Australia," observes Johnson, "we've only got one top predator left that's not a fox or a cat, and that's the dingo. We'd better make sure we manage it properly." Michael Soulé, who helped define the concept of mesopredator release, agrees. "Whatever damage the dingo has caused in Australia is already done," he says. "By my calculus it's clear that the dingo now creates a net benefit in the ecosystem."

Changing popular perspectives on the dingo is a challenge in a country where dingo management has long meant shooting or poisoning as many of the animals as possible. In Australia, dingoes exist in a strange conservation limbo. Depending on whom you ask or which law you consult, they are alternately described as native or introduced, as pestiferous vermin or an integral part of the country's ecosystem. In the Northern Territory, dingoes have been protected for some time, and the state of Victoria recently listed them as a threatened species. However, in most other states, rural landowners are legally required to kill dingoes on their property. Even in national parks, where the conservation benefits of dingoes are starting to be recognized, officials poison-bait large buffer zones to keep dingoes from straying onto neighboring grazing lands.

Under the Threatened Species Conservation Act of 1995, any animal that inhabited Australia in 1788, at the time the first British colony was established, is eligible for protection. The dingo was nominated for protection in New South Wales more than a decade ago, but the state government never acted on that petition. The state has, however, accepted a separate petition that defines hybridization between feral domestic dogs and dingoes as a "key threatening process" affecting dingoes. The finding is likely to lead to a continuation of the state's long-standing practice of air-dropping poisoned baits across much of the landscape. The nomination, explains ecologist Michael Letnic of the University of Sydney, effectively declares that there are few, if any, true dingoes in New South Wales, due to interbreeding with feral dogs.

For some biologists, the prospect of genetic pollution from free-roaming domestic dogs is a major threat facing "pure" dingoes. For others, the issue is overblown, too often used as justification for mass predator-control campaigns. Asked to explain the difference between dingoes and wild dogs, Letnic nods toward the ginger-colored pelt of a dingo lying across his office chair. "That," he says, "is a wild dog. It's difficult because basically dingoes are dogs; they are an ancient domestic dog. Seventy percent of the animals in the wild look and act like dingoes, but if you call them wild dogs you can throw poison out the back of a helicopter and that's fine." O'Neill and Wallach also see the hybridization problem as unimportant compared to the need to protect Australia's last top predator. "The best way to overcome it is just to leave them alone, stop baiting them," says O'Neill. "Gradually the domestic dog genes will be winnowed out through the process of evolution. Dingoes are going to be selected for over domestic dogs because they can survive better in the wild."

Letnic has been comparing the ecologies of outback habitats on opposite sides of the dingo fence. His results are still emerging, as he grapples with the logistical challenges of studying some of the most remote areas in Australia. But one trend stands out. "In a nutshell, there is more biodiversity where dingoes live," he says. Dingoes lower the numbers of foxes. They control kangaroo populations, which can boom in places where ranchers dig artesian wells to provide water for their stock. Where dingoes roam, native species—including rodents, lizards, and grasses—are more abundant and diverse.[21]

Turning to the dingo as a savior of biodiversity goes against ingrained ideas of what is, or is not, natural. Yet the much-persecuted dingo appears to be a successful, if unplanned, example of Pleistocene rewilding.

Protecting the species could yield real conservation benefits, without transplanting a single animal. Over the course of 4,000 years, the dingo has proved its mettle as a survivor and an efficient top predator. The first dingoes may have tumbled onto its shores straight out of human hands, but their descendants are now essential to the survival of Australia's wildness.

WILD BY DESIGN

New Zealand, lying 1,240 miles of open ocean away from Australia, was for most of its prehistory an utterly isolated world, a nation of bizarre and wonderful birds. Until the first humans arrived, 750 years ago, the only mammals that inhabited the islands were bats. Birds filled every imaginable niche, feeding on fruit, seeds, grass, shrubs, insects, and one another. Great ostrich-like moas lumbered over the landscape, chomping on branches. They would have towered over the early human settlers. The country was full of oversized and flightless birds, including rails, geese, ducks, and parrots. Raptors were the only threat the avian giants faced. The fearsome Haast's eagle, the largest predatory bird known to science, was so big that it existed at the physical limits of feathered flight and at times attacked even the largest of the moas.[1]

In some ways, New Zealand is a prime candidate for planned rewilding—advocates envision ostrich or emu standing in for moa in a reconstituted ecosystem. It is tempting to imagine such a resurrection in the country that hosted the world's most obvious, infamous, and recent case of prehistoric human overkill. The first inhabitants—Polynesians who would develop the unique Maori culture—reached the islands late in the thirteenth century. It took them only 100 years to exterminate a vast population of moa comprising ten different species.[2] Fifteen other kinds of native birds, including ducks, geese, pelicans, and coots, also vanished. Rats that had come ashore from the settler's canoes raided the nests of ground-dwelling birds. The Haast's eagle disappeared, as did an indigenous harrier—the island's new occupants did not leave them much to hunt.

Fig. 26   Pioneering paleontologist Sir Richard Owen with the skeleton of a giant moa. (Photo courtesy of Alexander Turnbull Library, Wellington, New Zealand.)

Those earliest people settled at river mouths, where they flourished on a bounty of moa and seal. Their village sites, uncovered by the plows of European settlers in the 1800s, contained enormous deposits of moa bones and eggshell fragments. Moa were so plentiful at first that the people discarded the heads and necks, eating only the choicest body meat. The archaeologist Atholl Anderson has estimated, based on the density and distribution of moa bones at known butchery sites, that hundreds of thousands of moa were killed and eaten.[3] His excavations at Shag River mouth show that a single settlement of moa hunters there slaughtered almost 9,000 adult birds. At first, moa and seal accounted for about 90 percent of the villagers' diet. Within twenty years, they had turned to fish for much of their sustenance, and the village disintegrated sometime in the fourteenth century, only a few decades after its establishment. By then, the local moa population was gone.

Fossil nests show that moa laid only one egg at a time. These were massive, typically about the size of a basketball. (Whole moa eggshells still exist, recovered from early Maori burial sites. People hollowed the eggs out, used them as water jugs, and laid them to earth in the arms of their dead). Before human arrival, the moa's only predators were hawks and eagles that attacked from the sky; they had no defense against the new hunters. Like other slow-breeding megafauna, moa populations were vulnerable to even a low-level take. As human numbers boomed, fueled by a steady harvest of moa meat, the birds quickly vanished.

Archaeologists have spent decades analyzing the details of early Maori culture, with its beautifully carved moa-bone fish hooks and ornaments. Only recently have ecologists begun to examine the details of the moa's life, in hopes of understanding the prehuman biology of New Zealand—and perhaps of finding an ecological stand-in. At first they depended on clues from scattered remains of moa that had died mired in swamps. Among their preserved bones paleontologists found the contents of the gizzard, the muscular pouch in the digestive tract that grinds up food, using stones the bird swallows. Moa were so big that their gizzard stones could be the size of a hen's egg. Mixed in with the rocks researchers found ancient twigs—bits of native shrubbery that constituted the moa's last meal.

The twigs in prehistoric gizzards showed that the moa fed on some strangely adapted trees unique to New Zealand. As young saplings, these trees sprout a complex interlacing of widely angled, wiry branches studded with tiny leaves. Once the plant grows taller than 7 or 8 feet—beyond the reach of a hungry moa—it shape-shifts, growing typical, straight branches with short twigs and large leaves. One example is the ribbonwood tree, though more than fifty other native plants show this eccentric growth pattern. In terms of capturing energy from sunlight, the juvenile form is terribly inefficient, but at the same time, it is very difficult for birds to eat. Scientists believe that these "divaricate" plants evolved alongside the moa, sacrificing photosynthetic capability in early life for protection against browsers. This costly defense has been rendered useless in modern times, however. Instead of moa, the plants now face introduced red deer, cattle, and goats, which can chew right through the complex net of young branches.

To test the idea that these New Zealand plants evolved their odd traits as a defense against moa, ecologists William Bond and William Lee fed samples to captive ostriches and emus.[4] These big birds, members of the ratite family, are among the closest living relatives of the moa. Offered wiry juvenile ribbonwood, the birds struggled to bite off the tiny leaves and break the resilient branches. Adult plants, on the other hand, made much easier pickings. The data suggest that ostriches and emus could survive on adult ribbonwood but would starve on the juvenile form.

Mammalian browsers like the goat use their tongue and lips to position food and their strong jaws to shear pieces off. A goat offered young ribbonwood branches was thus able to break off and devour four times the biomass the emus could manage. In Africa, land of the living megaherbivores, trees defend themselves by growing long thorns that stab giraffes and antelope in the mouth. By contrast, New Zealand has very few thorny plants. Moa had hard, heavy beaks, and thorns would not have bothered them. But the wiry structure of young divaricate plants likely slowed them down. Recent evidence from preserved gizzards and moa nests shows the feathered giants were capable of clipping half-inch-thick branches with powerful bills that acted like pruning shears.[5] The majority of the sticks they used came from the straighter, adult form of divaricate trees.

Bond and Lee estimate that ratite-resistant trees and shrubs are more than twice as likely to be threatened in New Zealand compared to other native plants. Introduced mammals, including goats and deer, eat right through their defenses and at the same time stomp on plants and disturb the soil, creating openings for exotic weeds. Although it is tempting to contemplate importing ostriches or emus as moa replacements to help disperse the seeds of native plants—a job deer and goats do not accomplish—most moa researchers are dubious about this prospect.

"Neither ostrich nor emu are great analogs for moa, unless we put them on steroids and run them through an intensive breeding program to get them used to living in forest," says Lee. The cassowary, a startling, blue-crested ratite from Australian forests, might make a better substitute, but it eats fruit instead of browsing on branches. It would take a combination of all three species to approximate the environmental impacts of moa, and even then, the imitation would be seriously flawed—none of the living ratites possess a powerful, pruning-shear bill.

Perhaps the clearest, and most surprising, clue to the void left by the moa's extinction was a discovery by Jamie Wood, a young paleoecologist who recently completed his doctoral work at Otago University in

Dunedin, New Zealand. While excavating plant fossils in caves and rock shelters along the southern end of New Zealand's South Island, Wood happened on a trove of preserved moa droppings. The mummified spoor revealed which plants the birds had eaten. Many of the coprolites contained seeds of tiny herbs that today are among the most endangered plants in the country.[6] It is hard to imagine giant moa bending their necks to the ground to feed on these dainty plants, but tests show that these herbs are packed with nutrients at concentrations far higher than in twigs and leaves of divaricate trees. Furthermore, they sprout only in the spring, when the moa likely nested and most needed a nutritional boost. The relationship was mutually beneficial—moa and other native birds were critical seed dispersers for these herbs.

Wood argues that there is no suitable ecological substitute for the moa. No mammal even comes close, and surviving ratite birds from Africa, Australia, and South America differ from moa in important ways. In his analysis of 1,200 moa coprolites, Wood notes that this new information on moa diet raises doubts as to how many suggested proxy species would make valid stand-ins for other extinct Pleistocene giants. As for the vanishing spring herbs, if introduced to the few spots where the plants now survive, native ducks and geese—or even introduced varieties, like the Canada goose—might disperse their seeds as effectively as moa once did.[7]

THE PACE OF THE TORTOISE

A fabulously weird kind of tree grows in a small tract of native forest on the island of Mauritius. Every year the trees sprout dazzling pink flowers at the base of their trunks: with time the flowers fade and fat green fruits encrust the bark. The plant, *Syzygium mamillatum*, has no common name and was not even known to science until 1987, though it lives in one of the world's most intensively studied ecosystems. The home island of this highly endangered plant, which lies east of Madagascar in the Indian Ocean, has become a sad icon of human-driven extinctions. Dennis Hansen, an ecologist at Stanford University, has a favorite line: "Everything that should not be done to an island has been done to Mauritius."

The first long-term human settlement was established by Dutch sailors in the mid-1600s. They found easy prey: the dodo, a 40-pound, flightless member of the pigeon family, along with endemic giant tortoises. Neither animal had experience of people and neither had any defense against them. Journals of the early settlers complain that while it was easy to get, dodo meat was tough and tasted bad. The fat-bodied dodo, with its thick hooked bill and quizzical expression, has become an international symbol of human-caused extinction. As is often the case on islands, its demise was likely driven less by human hunting than by the exotic predators the Dutch brought with them: rats, cats, pigs, dogs, and macaque monkeys, all of which raided dodo nests. By the close of the seventeenth century, only a few decades after it was first described in writing, the dodo was gone forever. Mauritius has also lost its tortoises, along with a giant skink that grew 2–3 feet long.

The fate of *S. mamillatum* was tied to those vanished island giants. The plump fruits that stud its lower trunk are designed to lure tortoises and skinks, which eat them and disperse the seeds within. Without these animals, the fruits rot on the ground. Most seeds that drop beneath the parent tree will not sprout, and the few that do seldom survive, shaded by the adult above them and vulnerable to fungi and insects. To thrive, seeds need to travel away from their source—and they need the large fruit eaters to carry them. The problem, explains Hansen, is that the only remaining frugivores on Mauritius are too small: tiny geckos or song-birds the size of a robin. A large endemic fruit bat is not much help, because it tends to nibble at the pulp and spit the seeds out. The seeds of *S. mamillatum*, like those of many large-fruited plants, are designed to journey through an animal gut before they germinate. Some seeds need a bit of friction to trigger them to sprout. In the case of others, the fruit pulp contains germination-inhibiting chemicals that can only be broken down by digestive enzymes.

Hansen and his colleagues fed *S. mamillatum* fruits to captive Aldabra tortoises, sorted through their droppings, and then planted the found seeds in a conservation reserve in Black Gorges National Park on Mauritius.[1] They discovered that seedlings from the tortoise-passed seeds grew taller and leafier than those grown from undigested seeds. Mauritius's own native tortoises are extinct, but living tortoises from other island chains, like the Aldabra, are similar enough to make good substitutes. Hansen hopes to see Aldabra tortoises introduced to Black Gorges in the next few years. The strategy is already working on the small, uninhabited Ile aux Aigrettes nearby, where free-ranging Aldabra tortoises subsist on the fruits of an endangered ebony tree. Before the tortoises were introduced, the trees were confined to a single small patch; now seedlings grow all over the islet.

Megafauna are scaled to the size of their home land mass. The conti-nents had mastodons weighing thousands of pounds, the large island of Madagascar had its 900-plus pound flightless elephant bird, and on Mauritius 220-pound tortoises were among the largest native herbivores. Regardless of scale, the biggest creatures in any habitat play a vital role.[2] "Without megafauna you have an empty forest, more of a botanical garden than a well-functioning ecosystem," says Hansen. We have lost many species—the discrete pieces that make up ecosystems—but worse, we have lost the interactions that make natural communities work. The main goal of ecological restoration, according to Hansen, should be to restore those relationships. All the requisite parts—plants, fruit eaters, grazers and browsers, predators, pollinators, and decomposers—fit together in a functioning, evolving system. If one part disappears (as did all the large frugivores in Mauritius), sometimes the only solution is to borrow spare parts from someplace else.

Jane Bock has strong reservations about inserting exotic spare parts into ecosystems. A botanist and ecologist who has devoted her life to the study of western deserts and grasslands, she can rattle off a litany of plants brought to the Americas with the intent of saving degraded cattle range, all of which have turned into uncontrollable invasives. Lehmann's lovegrass was imported from South Africa, but because it reproduces faster than native grasses, and cattle prefer the flavor of the natives, it has completely replaced some indigenous grass species. Buffelgrass, another African transplant, has invaded much of the desert land in northern Mexico and the southwestern United States. It thrives on frequent fires, rapidly reseeding and spreading, outcompeting cacti and other native plants that are not fire-adapted. "Introduced species often have no predators, or carry unique diseases that can wipe out native plants," says Bock. "Many introductions just corrupt the balance in nature."

Yet, despite their doubts, she and her husband, zoologist Carl Bock, signed on as coauthors of Josh Donlan's Pleistocene rewilding proposal. Their invitation to the Ladder Ranch conference was arranged by their old friend, Paul Martin. "I love Paul's vision of the megafauna and their impacts during the last Ice Age," says Bock. "I understand his longing to see that ecosystem resurrected." She does not, however, see this as practical in a West where open land is rapidly disappearing, morphing into a series of tidy ranchettes. Nevertheless, the conference allowed Jane to resolve the overpowering worry in her life at the time, the fate of a captive colony of endangered Bolson tortoises.

Under severe threat from hunting and habitat loss, the species survives only in a scrap of protected land in Mexico known as the Bolson de Mapimi. The Bolson tortoise has become a slow-moving flagship for Pleistocene rewilding. During the Ice Age, the tortoises roamed much of what is now the southwestern United States. Its fossils have been found as far north as the Grand Canyon in Arizona and as far eastward as Texas, Oklahoma, and Kansas.[3] Martin and others believe that late Pleistocene people hunted the species out of much of its original range. Whatever the case then, it is certain that people preying on the Mexican tortoises have contributed to their rapid decline in historic times.

The Bocks had worked for decades on an Arizona ranch owned by Ariel and Frank Appleton. During a stint as Peace Corps volunteers in Costa Rica, the Appletons had learned about the problems posed by over-grazing. On their return to the United States, they decided their 6,000 acres would be more useful as an ecological research center than as cattle range. In 1971, tortoise researcher David Morafka brought the Appletons a female Bolson tortoise from Mexico. Ariel named the tortoise Gertie and spent long days following her around, observing what the animal ate and how she dug her burrow. Over time, Ariel obtained more Bolson tortoises and coaxed them into breeding successfully (a feat many biologists and zookeepers had attempted, without success). By the turn of the twenty-first century, her tortoise colony had grown to thirty healthy animals, but Ariel was too ill to care for them anymore. She died in 2004, around the time Donlan and Greene began organizing their rewilding conference.

Jane Bock, who had been close with Ariel Appleton and her adopted tortoises, felt desperate to find a viable future for the captive Bolsons. On visits to Ted Turner's ranches in New Mexico, Bock noticed that the tortoises' preferred plants were abundant there. After the rewilding conference, the Bocks persuaded Joe Truett, an ecologist with the Turner Endangered Species Fund, to take on the Bolson colony. In September of 2006, twenty-six of the tortoises were released into 8-acre enclosures on the Armendaris Ranch in New Mexico.

Man-made burrows had been prepared for the tortoises, which rely on these shelters for protection from extremes of desert heat and as safe havens during winter hibernation. The animals settled in, using the provided burrows and excavating some new ones. Gertie—weighing in at twenty-five pounds and nearing seventy years old—was among them. She lives contentedly at the Armendaris Ranch, where, like the other tortoises, she feeds on native plants growing in the enclosure. She seems to enjoy booting smaller tortoises out of their burrows.

The most dangerous moment in a tortoise's life comes soon after it hatches, when it makes an easy target for predators. To build up the population, biologists remove some eggs from the enclosures and incubate them indoors, "head-starting" the hatchlings in a predator-free environment. The colony is thriving now, and the next step would be to release some of the animals on the Armendaris Ranch. This would require permits from the U.S. Fish and Wildlife Service. The Bolsons exist on U.S. soil today only because they were imported before they were listed as endangered in the United States and Mexico. Current U.S. law, as well as the Convention on International Trade in Endangered Species (CITES), prohibit bringing such animals across national borders, even for purposes of restoration.

Pleistocene rewilding has generated enthusiasm, scorn, and a great deal of media hubbub over the idea of lions and elephants loose in the American West. Now, even Donlan, rewilding's strongest proponent, sees the humble tortoise as the most realistic starting point. Donlan has worked with Dennis Hansen on a review of tortoise restoration projects world-wide. They lament the obstacles imposed by restrictions on the international transport of tortoises for any reason, including rewilding. It is time, they argue, to drop these barriers. For-profit trafficking in endangered species is one thing, but returning animals—or their analogs—to their prehistoric ranges is quite another.[4]

Any rewilding scenario is rife with potential glitches. Big herbivores will overpopulate in the absence of effective predators, but people have a long history of conflict with and intolerance toward large carnivores. A healthy community of reconstituted megafauna would need vast expanses of land, a space far bigger than Yellowstone. Donlan and Greene emphasize that they hope to start experiments with Pleistocene rewilding on a small scale, in fenced enclosures. Yet the history of fenced wildlife reserves, like South Africa's Kruger National Park, has already demonstrated that what elephants and other megaherbivores need most is room to move. Kruger, at about the same size as the state of Massachusetts, has chronic problems with elephant overpopulation.

Though the original rewilding proposal talked of lion, cheetah, and elephant, the Bolson tortoise is a good place to start: a creature that can motivate and inspire the public without causing panic. So far, the Bolson's return to the United States has been uncontested, perhaps because there is good fossil evidence that the same species lived here during the Pleistocene. By this standard, we should also be bringing jaguars back to

Southern California and Arizona, where they once roamed along with saber-toothed cats and camels. Instead, we build fences along the Mexican border, cutting jaguars off from the last available habitat in the United States.

Some scientists argue that the federal government quietly committed an act of Pleistocene rewilding in 1996, when it released endangered California condors at the Vermilion Cliffs in northern Arizona. Condors had patrolled the region during the Ice Age, carrying chunks of dead megafauna to their young sheltering in high, dry caves along the cliffs of the Grand Canyon. The Vermilion Cliffs release was justified on the basis of a few written records of nineteenth-century condor sightings in Arizona, Utah, Idaho, and Montana. But according to paleontologists, the great birds vanished from the continent's interior 13,000 years ago, along with the mammoth, ground sloth, camel, and horse.[5] The birds recorded in the 1800s represented a brief blip of population reexpansion, made possible by the livestock brought to the region by Euro-American settlers.

The condor makes a fascinating case study of human-driven extirpation and rewilding. The species has survived such cycles more than once. Giant vultures that rely on carrion for sustenance, condors stand close to 5 feet tall and soar on wings that span more than 9 feet. With their naked red heads and black plumage, they are not pretty. But even with conspicuous radio-tags on their wings, free-roaming condors convey the wild spirit of the Pleistocene. The birds flew over much of coastal and southern North America during the Ice Age and lived as far east as Florida and New York. After the megafauna extinctions, food became scarce, and they survived only along the Pacific coast, where they fed on the washed-up corpses of fish, whales, and sea lions. Lewis and Clark watched a flock of condors feeding on a dead whale at the mouth of the Columbia River in 1806; the birds remained common along the California coast until the mid-1800s.[6] By then, commercial whaling and the fur trade were devastating marine mammal populations, depriving coastal condors of food. At the same time, European settlers were populating the landscape with livestock, whose carcasses helped feed condor populations for several decades, though the birds were often shot by ranchers or for sport. Later, as fruit groves and vegetable farms replaced cattle and sheep ranches, condors again lost their food base. The slow-breeding creatures (they do not become sexually mature until age six, and adult females lay only one egg every other year) dwindled under this onslaught of troubles. By the 1940s, the species' range had shrunk to a small band of habitat in the mountains north and west of Los Angeles.[7]

Condor numbers continued to drop: many birds died of lead poisoning from eating the carcasses of deer contaminated with lead shot. In 1986, the last wild birds were captured to be used in a breeding program thought to be the species' last hope. Some warned that attempts to rescue the condor were doomed: without a steady source of uncontaminated carrion, the birds would always need to be fed. Captive-bred condors—and their offspring—now fly free in Arizona, California, and Baja, living on carrion they find themselves supplemented by safe, lead-free calf carcasses provided by wildlife management agencies. No population survives in the wild without supplemental feeding, although some birds prefer to find their own food and are moving toward self-reliance.[8]

Condor repopulation efforts along the Pacific coast face different obstacles. Birds at Big Sur on the Southern California coast have begun to feed on seals and whales, but along with the meat they ingest the pesticide DDT, which concentrates in the blubber of marine mammals. The poison causes condors to lay thin-shelled eggs, which break before the chick inside matures. For now, biologists replace the eggs laid by condors living near the coast with those laid by captive birds. Some activists, including the Yurok of Northern California, a tribe that uses condor feathers in its ceremonies, hope to bring the big birds back to less contaminated stretches of the Pacific coast. First, however, they must prove that local food sources are clean enough to sustain healthy condors. Condors may never make it in a human world, until we find a way to radically change—cleanse the oceans of DDT, eliminate lead shot, or fill our open spaces with imported megafauna.

Still, the condor restoration program shows that conservationists can overcome political barriers to rewilding Pleistocene species if, like the tortoise and the condor, they appear nonthreatening. Meanwhile, the renewal of historic native species like wolves, prairie dogs, black-footed ferrets, and bison remains contentious. The pitch for Pleistocene rewilding can be viewed as a mere exercise in the rattling of ideological cages. Yet lurking behind all the fuss is an idea that can open pragmatic new perspectives on conservation. Despite its obvious problems, the faithful believe it is possible to start small and build upward. They may be hopeless dreamers—or they may just be that small, committed group of people who succeed in changing the world.

Rewilding advocates often base their arguments on human responsibility for mass extinctions, whether they focus on the ancient losses of native giants in the Americas and Australia or the relatively recent dieoffs in New Zealand and Madagascar. Now, however, even for devotees of Martin's overkill theory, climate has become a crucial issue. The warming trend people have triggered through the burning of fossil fuels is more extreme, and is happening faster, than any of the Pleistocene warm spells. The creatures most threatened by global warming are rare Ice Age survivors, who soon may have no place cold enough to offer them a safe harbor.

THE
BIG HEAT

MELTING ICE

Slate-colored waves slam onto the beach at Point Barrow, Alaska, the northernmost sliver of land in the United States. To get here from the town of Barrow, Dan Lum bounced several miles over washboard ridges of beach gravel in his battered van, its seats and body cracked from the subzero cold of Arctic winter. As he drove along, Lum, a big, genial man with his shirt sleeves rolled up in the 40-degree summer air, passed ceremonial sites used by his Inupiat ancestors for millennia, now being eaten away by the sea. Until a few years ago, sea ice kept the waters calm even in high summer. Now the ice has retreated to 200 miles offshore, and the ocean roils under strong winds, eroding this narrow strip dividing the Chukchi and Beaufort seas.

Barrow occupies ground that once held the ancient Inupiat whaling village of Utqiagvik. From the edge of town, you can see for miles across the tundra, glimpsing snowy owls, white-fronted geese, and caribou—the once-typical Arctic fog has vanished along with the ice. For natives like Lum, this weather feels as strange as recurring blizzards would to residents of Los Angeles.

The beach at Point Barrow is laden with fascinating flotsam: the knuckle bone of a walrus, a seal scapula, and scattered slivers of mammoth ivory. Thousands of years after their extinction, preserved fragments of Arctic elephant continue to turn up. In the era of man-made climate change, surviving megafauna face troubles that evoke the late Pleistocene. Some of the most threatened creatures come from here in the far north, the ancient home of the woolly mammoth, where air temperatures are rising at twice the global rate. The Inupiat settled here long after the last mammoths had died out but have nonetheless always relied on big mammals: bowhead whales, walrus, and caribou. Humans are relatively recent arrivals in a community of megafauna adapted to the stark landscape of the Arctic.

Lum packs a rifle, a standard precaution when venturing outside of Barrow; meeting a polar bear unprepared could prove fatal. But this summer, with the sea ice—the bear's hunting ground—so far off, the gun seems like a forlorn relic of colder times. Lum describes the now-distant ice, a tumbled collage of blue-tinged forms named in intimate detail in the Inupiaq language. The native people here are connoisseurs of sea ice, having used it for generations as a platform from which to hunt bowhead whales, their main source of sustenance. The shifting location and thickness of sea ice can spell the difference between hunger and plenty for the hunters of marine mammals in the far north—whether human or bear.

The polar bear evolved with its staple prey, the ringed seal, which swims below Arctic pack ice. The Inupiat have likewise been defined by their relationship to the bowhead whale. More than 1,500 years ago, the coastal people of Alaska's north slope mastered the technique of hunting whales, launching their small, light sealskin boats from camps at the edge of the shore-fast ice in spring and fall, sinking stone and antler harpoons into the ocean giants' thick blubber. The Inupiat, for their part, have successfully weathered past warm spells. Around 800 c.e., the northern hemisphere heated up as a result of natural climate cycles. Glaciers shrank in Europe and North America, and the sea ice drew back from shore. The Medieval Warm Period lasted about 400 years, shaping human history in Europe, where the Norse colonized Iceland and Greenland, and the British reaped rich harvests and took up wine making. At the same time, the retreat of sea ice created leads, or stretches of open water, all along the Arctic coast, places where whales could reliably be taken using strategies honed on Alaska's north slope.[1] A single family, hunting from two or three kayaks, could take a whale that yielded enough meat to feed them for months. The whaling lifeway spread quickly across the breadth of Canada and into Greenland. This pan-Arctic whaling tradition—labeled as the Thule culture by anthropologists who first encountered it in Greenland—thrived until the climate shifted back into a colder phase, around the year 1300.

That cold snap, dubbed the Little Ice Age, changed life for people throughout the far north. Farmers in Iceland could no longer grow a successful crop of barley. Sea ice off the Arctic coast thickened, and nearshore leads, so useful for whale-hunting, disappeared. On the coast of central Canada, whaling people gave up permanent land dwellings to spend winters on the pack ice, living in igloos and hunting seals—a human variation of the polar bear's strategy.

Glenn Sheehan, an anthropologist who has spent decades in Barrow studying Inupiat society, explains the dilemma that Alaska's Inupiat faced at the dawn of the Little Ice Age. They could have continued living in independent, scattered groups along the coast, forsaking their bond with the bowhead whale and adapting to a diet of seals, caribou, and fish. Instead, they chose to gather at the few points of land where leads reliably formed close enough to shore that the enormous mass of whale meat could be transported back home, to villages like Barrow, Wainwright, and Point Hope. Staying faithful to the whale meant living in crowded villages—and finding ways to keep the peace. When the coastal ice had been pocked with open leads, anyone with a grievance against his neighbor had the option of picking up and moving farther along the shore. In colder times, working together became essential.

The Inupiat calendar is built around the bowhead's movements. The whales winter between Russia and Alaska, in the Bering Sea. In the spring, they migrate along Alaska's western coast to the Beaufort Sea, traveling through leads where the shore-fast ice shears away from the polar ice pack.[2] In autumn, they return. For the Inupiat, spring and fall are whaling seasons, which hunters spend out on the ice. In a state of constant vigilance, they watch not only for passing whales but also for subtle movements of the ice: heavy chunks of pack can drift, colliding with and crushing a whaling camp in moments. In summer they hunt caribou and waterfowl. The long dark winters are devoted to making and mending kayaks and other whaling equipment. Modern Inupiat use snowmobiles and explosive harpoon tips, but the cycle of their seasons remains the same.

Unlike most other baleen whales, which migrate to the tropics in winter, bowheads spend their entire lives in icy northern waters. For insulation, they grow a thick layer of blubber that once made them a prime target for commercial whalers; a bowhead renders more oil than any other whale. Europeans first hunted them off the coast of Greenland, as early as the 1500s. Yankee whalers discovered the Bering-Chukchi-Beaufort (BCB) sea bowhead stock—the lifeblood of the Inupiat—in the nineteenth century. A booming market for baleen, used to make Victorian ladies' corsets, buggy whips, and even early type-writers, had made the hunt more profitable than ever. But bowheads possess the traits that make megafauna so vulnerable to overkill: they are big and obvious, they live long (often more than 100 years), and repro-duce slowly. Before commercial whaling, bowheads throughout the Arctic numbered at least 50,000. By 1910, when the whaling frenzy began to slow, the world population had been reduced to a few thousand animals.[3]

The Inupiat were hit as hard as the whales. Many died of hunger as whalers devastated their traditional food sources, taking not only whale but all available walrus and seal; others succumbed to European diseases, like measles, to which they had no immunity. Still, the native culture managed to survive, and today Inupiat life remains centered on whales. (The carefully managed subsistence hunt by native people takes about 0.5% of the bowhead population annually.) Meanwhile the BCB stock, in its relatively undisturbed habitat, is recovering quickly. More than 10,000 whales migrate past Barrow each spring, and the population increases by about 3.4% each year.[4]

No one knows yet how the bowheads will be affected by warming and the retreat of the ice. Some evidence has shown that less ice cover may mean more food for bowheads. When they feed, they filter small, drift-ing sea creatures through their baleen. Copepods, tiny translucent beasts that resemble miniaturized lobsters, are a staple of their diet. In some parts of bowhead habitat, copepod numbers seem to be booming as the melting ice lets more sunlight into the water, fueling growth of the algae that form the base of the food chain, and opening new feeding areas for the whales.[5]

Even if whales thrive as the pack shrinks, the state of the ice is of criti-cal importance to the Inupiat hunt. In the spring of 2009, says Sheehan, native hunters managed to take four animals in particularly difficult and dangerous conditions, working on thin ice and impeded by high winds. As the ice continues to grow thinner and retreat farther from shore, hunt-ers may no longer find surfaces solid enough to hold a whale carcass.

They may be forced to work with ice weaker than any encountered in Inupiat history. But for now, the whaling tradition goes on.

The shared need for a stable platform of sea ice forms a powerful bond between the Inupiat and their fellow hunter, the polar bear. Native people around Barrow have always hunted bears, sometimes falling prey to them in turn. Inupiat archaeological sites hold varied polar bear relics: mittens from their hide, amulets made of their teeth, tools carved from their bones. The Inupiat have a unique understanding of the bear's dilemma as warming alters the face of the north. Yet they were unhappy when, in May 2008, the bear became the first species listed as threatened due to the impacts of climate change. Barrow locals worry that the federal government will restrict native hunting in the name of polar bear conservation. That action would not make a real difference to the bears' survival, but Eskimo subsistence hunts are far easier to regulate than the root problem of greenhouse gas emissions.

On the remote pack ice of the Beaufort Sea, to the north and east of Barrow, researchers have been finding evidence of global warming's toll on polar bears. Normally the bears hunt by waiting patiently at ringed seal breathing holes, or by breaking through a thin crust of snow hiding the lairs where mother seals nurse their pups. As the ice pack shrinks, some bears have tried to claw through solid ice in a desperate effort to capture seals—a tactic with little chance of success. In recent years, researchers have witnessed other strange events, unlike any recorded in three decades of scientific bear watching. Bears are turning cannibal, the largest males feeding on smaller females and yearlings. During a helicopter survey in January 2004, Steven Amstrup of the U.S. Geological Survey discovered a polar bear den that had been broken open. Following a trail of blood, he found the still-warm body of a female bear, killed by a big male whose stabbing canine teeth had penetrated into her braincase.[6] The killer had attacked the female as he would have a ringed seal—with a devastating bite to the head.

Among polar bears, only pregnant females hibernate. Males and youngsters stay active, hunting through the winter. Expectant mothers shelter for the winter in snow caves near the coast or on drifting pack ice, where they birth their cubs and raise them for the first three months of their lives. Females prefer denning away from the places where male bears tend to hunt. The slaughtered female was found outside her den on Pingok Island in the south Beaufort Sea, far from the area where male bears usually forage in winter. The evidence suggests that the male, desperate for food, had come to Pingok specifically to hunt denning female bears; there was no other reliable source of food in the area.

In the summer of 2005, sea ice cover in Arctic waters shrank to the lowest extent ever measured in nearly thirty years of satellite monitoring. The ice pack shriveled even more in 2007 and 2008. During the colder months, polar bears typically hunt at the seam between shore-fast ice and the long-lived pack ice that floats farther out to sea. When summer comes, the ice melts first along the heat-absorbing Alaskan coast, forcing them to make a choice. Most bears stay on the pack as it moves farther north. But the best seal habitat is in the shallow, fertile waters of the coastal shelf. The pack now retreats as much as 400 miles from the coast, leaving polar bears in the south Beaufort Sea with a choice between summering over deep ocean water where prey is scarce, or trying to survive onshore. So far, the bear population there is holding steady, but the animals grow thin and hungry in summers of little sea ice, and some perish when they face the winter without fat reserves. Bears living on the northern Beaufort Sea, where the ice pack remains more stable, are stronger and healthier and so far show none of the desperate behaviors that have occurred in the southern part of the sea.[7]

Meanwhile, studies in Canada's western Hudson Bay, at the southern limit of the polar bear's range, serve as a harbinger of things to come. The population there is in decline, the number of bears in the region dropping from 1,194 in 1987 to 935 in 2004.[8] Young and old bears are most vulnerable to the impacts of warming and reduced ice pack. Their survival rate correlates with the date of spring sea ice breakup in a given year; in 2004 the ice broke three weeks earlier than it had in 1987. Without any ice to hunt from, bears venture into the town of Churchill, foraging at garbage dumps and clashing with local people. Even so, the food scraps left by humans are no substitute for the rich nutrition provided by blubbery seals. Hudson Bay bears lose an average of about 1 kilogram (2.2 pounds) of body weight for every day they wait onshore for the sea ice to refreeze.

Climate models predict that in the foreseeable future, Arctic temperatures will reach highs unprecedented in the polar bear's

Fig. 27   A polar bear feeds on a whale carcass in the Arctic National Wildlife Refuge. Polar bears, bowhead whales, and Inupiat hunters are all powerfully affected by climate change in the Arctic. (Photo from Wikimedia Commons, http://en.wikipedia.org/wiki/File:Polar_Bear_ANWR_10.jpg.)

evolutionary history.[9] The species evolved from brown bear ancestors about 300,000 years ago, during a chill peak in Pleistocene glaciation. Moving from land to sea ice, they filled a new niche, harvesting the rich bounty of the ocean. On a diet of seals, they grew massive. Polar bears are the largest living members of the bear family, though their Pleistocene ancestors were even bigger. Males recently captured on the Beaufort Sea stood nearly 13 feet tall on their hind legs and weighed about 1,760 pounds. By contrast, Alaskan brown bears living adjacent to the sea ice are the smallest of their kind and occur at lower densities than brown bears anywhere else. The pickings in the terrestrial Arctic are far too slim to support polar bears. Their fate is tied to sea ice and to the seals beneath it. If the ice continues to shrink at the predicted rate, only one-third of the world's polar bears will survive by the end of the twenty-first century. Like the mammoths and giant deer of the Pleistocene, they will struggle to hang on in a fragment of their earlier range.

The polar bear tends to grab the headlines. But the reindeer, a creature that has fed and clothed people since they first spread out of Africa into Europe and northern Asia, may have more to reveal about climate-related changes past and those to come.

Reindeer were a mainstay of Pleistocene people of the north. During peaks of glaciation, the reindeer was commonly found as far south as Italy and Spain. A record spanning nearly 50,000 years of the interconnected lives of people and reindeer has been preserved at the Grotte archaeological site, a rock shelter on the Dordogne River in southwestern France. Donald Grayson, Paul Martin's long-time opponent in the overkill debate, has studied the site with his colleague Francoise Delpech. As the archaeologists dug deeper, reindeer became more dominant in the bone collections, all of which had been left behind by human hunters. Using a reconstruction of Pleistocene climate based on fossil pollen records from the region, Grayson and Delpech matched the relative abundance of reindeer with colder summer temperatures. The species thrived far to the south of its current range 40,000 years ago, when summers were 5.4–9 degrees Fahrenheit colder than in modern times.[10] The reindeer record at Grotte ends about 10,000 years ago, when summer temperatures in southern France climbed past the animals' tolerance limit.

This ancient pattern of boom and bust resonates with the findings of biologists studying modern reindeer. Of forty-three major herds monitored worldwide over the last decade, at least thirty-four are in decline, showing an average drop of 57 percent from historic population peaks. The trend applies to Old World reindeer and New World caribou alike (both varieties of the same species, *Rangifer tarandus*). Populations have always fluctuated with local shifts in weather and predation pressure. But in early 2009, Liv Vors and Mark Boyce of the University of Alberta reported the first evidence of a synchronized global decline in reindeer numbers.[11] The most obvious causes are linked to global warming.

Caribou walk a long tightrope between starvation and survival, migrating as far as 620 miles to reach their summer calving grounds. In spring they leave the shelter of boreal forest, which protects them from the brutal winters, and head to open coastal plains to birth their calves.

Timing is critical: females must bulk up on fresh plant growth in order to raise healthy young, as well as to build the fat reserves they need to survive another winter themselves. Rising temperatures in the Arctic now trigger plants to sprout earlier in the spring, so the animals have begun to slide out of sync with their summer food sources. In West Greenland, the growing season for important forage plants moved up by almost two weeks between 1993 and 2006. Caribou mothers rely on tender buds of dwarf willow and birch and the freshest shoots of grass and flowering plants. Soon after these spring plants sprout, they begin to lose their nutritional value. The caribou's urge to migrate is triggered by changes in day length rather than rising temperatures, so the herds have paid a heavy price. As the mismatch between the green-up and the arrival of migrating caribou increases, the death rate of calves rises while the body condition of adults deteriorates.[12]

In summer, when uninterrupted, peaceful grazing can mean the difference between life and death in the winter to come, caribou are plagued by parasitic insects that can stop them from feeding. The worst culprits are warble flies and nasal bot flies, which specialize in attacking the caribou. Warble flies lay their eggs on reindeer fur. When the larvae hatch out, they burrow under the animal's skin. Nose bot flies spray their larvae onto the muzzle, and they later crawl into the reindeer's nose and sinuses; a heavy infestation can obstruct the caribou's breathing. Warble and bot flies like warm weather and sunshine, and the level of insect harassment seems to rise with summer temperatures. To escape, caribou move to windy ridges or snow patches where there is little food. They will plunge their faces underwater to keep bot flies off. Sometimes the torment of the flies keeps the animals running all day long. After a hot summer of heavy fly attacks, both adults and calves are dangerously thin and stand little chance of surviving the winter.[13]

Parasitic flies afflicted reindeer of the deep past. Paleolithic artists in Europe carved and painted images not only of the reindeer they hunted but also of the warble fly larvae that infested them.[14] In 1997, researchers stumbled on ancient deposits of caribou dung in the melting remains of what had long been a permanent snow patch in southern Yukon, Canada. Among the droppings was a wooden dart, used about 4,000 years ago by hunters who knew caribou would be sheltering on the snow patch during

Fig. 28   Caribou populations are in decline throughout the far north. Warming temperatures cause early spring green-up in the Arctic, so migratory herds fall out of sync with critical food sources. Warming also favors parasitic flies, which take a heavy toll on caribou. (Photo courtesy of U.S. Fish and Wildlife Service.)

the heat of the day, trying to escape bot flies.[15] Now, as human-driven warming intensifies the bot fly problem, we may be witnessing a replay of one of the elements that drove reindeer out of their southern ranges in France and Mississippi at the close of the Ice Age. But surviving populations already live in the far reaches of the north; as the planet heats, there is nowhere left for them to go.

MOVING WITH THE TIMES

Wildlife around the planet evolved with climate change. During the shift from the last glacial maximum to the warm spell that has defined the Holocene epoch, global mean temperatures rose at about 9 degrees Fahrenheit per 5,000 years, or 0.0018 degrees per year. During the Medieval Warm Period, the heat wave that allowed the expansion of the Thule whaling culture 1,000 years ago, the rate of warming was about 1.8 degrees Fahrenheit per 100 years, or 0.018 degree per year. For this century, current projections from the Intergovernmental Panel on Climate Change forecast a rate of warming 10 percent faster than recorded for the Medieval Warm Period—in their best-case scenario.[1] It is more likely that temperatures will climb 100–300 percent faster. This rate of change is more rapid than any in the evolutionary history of mammals.[2] Worse yet, this intense hot flash begins in an already warm world, rather than a glacial maximum with much of Canada, Alaska, and northern Europe locked under ice.

During the Pleistocene, animals seeking better habitat as the climate shifted could roam freely across, and sometimes even between, continents. This time around, humanity has altered, paved, or fenced great swaths of the earth. Together, the rapid pace of warming and habitat loss are pushing wildlife to the wall.

A dramatic example is the case of the pika, an alpine member of the rabbit family. Pikas live on rock-strewn mountain slopes, issuing high-pitched whistles to declare their territorial rights and warn one another of danger. In summer, they gather leaves and flowers, which they store to sustain them during winter. With thick pelts and a normal body temperature of 104 degrees Fahrenheit, pikas can overheat quickly in the summer sun; they scamper in and out of the shade of rock crevices to cool themselves. The charismatic little creatures, which can often be seen carrying hay to their caches, are favorites among mountain hikers.

Biologist Erik Beever went searching the mountains of Nevada, Utah, Oregon, and California for historically recorded pika populations to gauge their response to warming. During fieldwork in the late 1990s, Beever found that seven of twenty-five pika populations recorded earlier in the twentieth century had vanished, an extirpation rate of about 30 percent.[3] Surviving populations had moved upslope, into higher, colder habitats. In a separate study, pikas in Yosemite National Park have been found to move 1,700 feet higher in elevation over the past 90 years.[4] As temperatures continue to climb, pikas will be increasingly isolated on small islands of alpine rock, unable to cross the warm valleys that separate one population from the next—a recipe for continued local extinctions.

American pikas weathered transitions from wet, cold glacial peaks to warm dry interglacials over hundreds of thousands of years of Pleistocene time. We know this because of a rare trove of Ice Age fossils found at Porcupine Cave, 9,000 feet high in the Colorado Rockies, which holds ancient pack rat middens, a library of bones that date from 600,000 to nearly one million calendar years bp. Through the ages, predators—including wolves, American cheetahs, bears, and cougars—dragged their prey there for a quiet meal, adding bones to the collection. Anthony Barnosky, a vertebrate paleontologist at University of California–Berkeley, spent fifteen years excavating at Porcupine Cave, seeing in the fossils a unique opportunity to explore the differences between the time of the mass megafauna extinctions and earlier interglacials set in a pre-human America. Barnosky began his scientific career as an overkill skeptic, convinced that climate change alone had driven mammoths and other giants to extinction. The evidence he gathered at Porcupine Cave helped to reshape his vision of humanity and nature in both the Pleistocene and modern worlds.[5]

Throughout the Ice Age, as glaciers retreated and advanced, the landscape outside Porcupine Cave shifted from dry and relatively warm to snowbound, damp, and intensely cold. Many of the smaller creatures living in the area were affected: one species of rabbit, mouse, or shrew might decrease or disappear, while another would become more abundant.

But overall, the animal community remained stable, and large species were seldom touched. For Barnosky, Porcupine Cave made the same point that Paul Martin had begun to argue two decades earlier. "You don't see big changes in populations of large mammals," he says, "until the mass extinctions at the end of the Pleistocene." In Colorado, and throughout much of North America, that moment coincides with the dawn of Clovis hunting culture.

The mammoth, mastodon, and ground sloth died out at a unique point in the earth's history, when human populations were growing dramatically and temperatures were climbing fast. The two threats combined to create a synergy of destruction, killing off large animals in numbers many times greater than had ever occurred in earlier interglacials. Barnosky sees the situation today as the same story, magnified. The current rate of warming is unprecedented, and human populations, already vast and continuing to expand, have blocked animals from moving around the landscape as they seek to find shifting patches of usable habitat.

Barnosky advocates the creation of habitat corridors to allow wildlife to move through the intricate web of man-made obstacles—cities, roads, fences, and farms—that covers much of the planet. The idea is based on decades of research in conservation biology, but it is difficult to put into practice as human populations grow around major wildlife reserves. In Africa, the continent that retains the most diverse array of megafauna, populations of big herbivores such as kudu, waterbuck, and roan and sable antelope are dwindling as increasingly hot, dry conditions deprive them of forage. A study of projected climate change impacts predicts South Africa's Kruger National Park may lose up to 66 percent of its native species to warming by the year 2050.[6] An analysis of large and medium-size mammals in reserves throughout Africa suggests that climate change alone—without accounting for the accelerating impacts of human habitat destruction—will bring 25–40 percent of these species to the brink of extinction by the late twenty-first century.[7]

A study of more than 1,500 native plants, insects, and birds in the Northern Hemisphere reveals that many populations are moving north at about 3.7 miles per decade, and up mountain slopes at about 20 feet in the same amount of time.[8] Biological events triggered by rising spring temperatures—such as the sprouting and blossoming of plants, the emergence of insects, and the migration of birds—are happening earlier. But as in the case of caribou and tundra plants, long-time ecological partners can become separated in time, with potentially devastating results.

It may be unrealistic to bet the future of the planet's wild plants and animals on our ability to protect effective habitat corridors in the right places, in a landscape that will be ever-changing as the climate warms. Some researchers have suggested that since we humans are in the way, we have a responsibility to start helping organisms to move poleward, or upslope, when they are not able to do it on their own. Many of the architects of this concept, called "assisted migration," study butterflies and plants, but the idea has important implications for modern megafauna.

Jessica Hellmann, now an ecologist at the University of Notre Dame, began studying the response of butterflies to climate change as a graduate student, more than a decade ago. She worked with the bay checkerspot butterfly, whose wings resemble a fanned patchwork of deep brown, bright orange, and white. The subspecies was once common in native grasslands around San Francisco Bay. Hellmann built on a foundation of data collected by the famed ecologist Paul Ehrlich, who had been studying checkerspot butterflies at Stanford University's Jasper Ridge Biological Reserve since the 1960s.[9] Over the decades much of the checkerspot's habitat has been paved over and built up, and the butterfly was listed as threatened in 1987.

Mammals can regulate their own body temperature, but insects' lives are governed by air temperatures: if it is too hot or too cold, adults cannot fly in search of mates, and if larvae fall out of sync with their host plants due to shifting weather, a generation may starve. It is well documented that butterfly populations expand or decline in direct response to changing temperatures or amounts of rainfall. In Hellmann's eyes, the bay checkerspot was the ideal canary in the coal mine of global warming: rising temperatures would be expressed among butterflies before they would show in populations of mammals or birds. A detailed knowledge of the creature's life cycle already existed, a rare advantage for researchers studying insects. Butterflies, explains Hellmann, make great study subjects because their beauty has attracted passionate study for decades.

Little is known of the natural history of most other insects, except for agricultural pests and disease-carrying mosquitoes.

The last two bay checkerspot populations on Jasper Ridge died out in the 1990s, an outcome made inevitable by fragmentation of the butterfly's habitat and speeded up by the impacts of climate change. Checkerspot larvae feed on only two kinds of native plants, plantain and owl's clover. Larvae hatch from eggs in spring, feed voraciously, and then enter diapause, a hibernation in which they wait out the heat of the summer. Unless enough fresh, nutritious fodder is available, larvae cannot grow enough to survive diapause. In times of drought, winged, adult checkerspots are found only on the coolest slopes; in warmer places plants shrivel in the sun before the larvae can bulk up.[10] In wetter years, more larvae survive. Hellman's work helped demonstrate that a dramatic increase in the variation of annual rainfall drove the Jasper Ridge butterflies into extinction. This increase in variability—with more years of extreme drought—fits with scenarios projected by global climate models.

The Jasper Ridge checkerspot populations blinked out in a sequence determined by climate. The group that lived on a flat atop the ridge vanished first, in 1991. The one that hung on until 1998 occupied hilly terrain, where cool north-facing slopes offered shelter from the sun. A few populations still remain, scattered on hills around San Francisco Bay, increasingly isolated from one another. The bay checkerspot is not a traveler, and the outlook for its survival is grim. Even when butterflies still lived in those last two patches on Jasper Ridge, only 550 yards apart, adults from the two populations rarely moved far enough to interbreed.

Having witnessed the fade-out of the bay checkerspot firsthand, Hellmann has pioneered the concept of assisted migration. She is one of the leaders of a group of scientists studying the issue and its legal and ethical ramifications. The concept is not as radical as Pleistocene rewilding. Still, any vision of biologists intentionally moving species beyond their historic ranges violates a long-standing conservation taboo. A prime candidate for assisted migration, for instance, is the endangered quino checkerspot butterfly, a relative of the bay checkerspot, now stranded in overheated habitat south of Los Angeles. This subspecies could likely be rescued by moving it past the sprawling city and into the mountains. It is easy and cheap to transport cocoons, and checkerspot populations rarely boom to the point of harming their plant hosts. At the moment, however, it is a violation of the federal Endangered Species Act (ESA) to move a listed organism outside of its historic range—regardless of the reason.

Like many environmental laws, the ESA is meant to preserve nature in an idealized, pristine state. That goal is out of sync with current knowledge of ecosystems shifting under the pressure of human impacts. Nature has never been static, but as Hellmann points out, climate change makes it particularly capricious. Realistic conservation will mean letting go of the idea of a stable state and learning how to protect nature in an ever-changing world. Even if one buys into the idea that conservation in North America should strive to recreate the world of 1492, that goal will become only more elusive as the planet heats up.

Humanity has a long track record of moving plants and animals to new environments where they become invasive, multiplying out of control and outcompeting natives, and many biologists worry that assisted migration would inadvertently create ecological monsters. Hellmann and her graduate student Jillian Mueller have analyzed the origins of 468 invasive species in North America.[11] They found that most of the problem organisms were carried here from other continents. But about 15 percent of the invaders were native to some part of North America and had spread to new regions with human help. One example is the rusty crayfish, carried by people from its home in the Ohio River to lakes and streams in the Upper Midwest and Northeast, where it knocks down native populations of plants and invertebrates. Another is the cowbird, which lays its eggs in the nests of other songbirds, leaving them for smaller birds to raise. Found only in prairies before European settlement, the cowbird has expanded its range in the wake of mass deforestation and now parasitizes the nests of more than 220 bird species.

Hellmann's study of intracontinental invaders offers a glimpse of what might happen with assisted migration. A small number of species likely will go invasive, but the risk varies among different organisms. Freshwater creatures, like the rusty crayfish, have a much higher probability of running amok than do terrestrial ones. This kind of information, along with studies of each candidate for relocation, will be needed to perform a kind of triage to decide which species might benefit from assisted migration without causing serious harm.

Assisted migration would inevitably be a balancing act between different kinds of risk. There is a chance that we could create a new invasive species; there is also a chance that we could waste precious members of endangered populations in relocation attempts that ultimately fail. But doing nothing also holds dangers. Some populations or species may go extinct; economic and ecological benefits could be lost if, for example, we fail to start planting key tree species north of their historic ranges.

Moving any species beyond its historic range will require the same kind of advocacy that returned the wolf to Yellowstone after decades of fiery debate, and only charismatic creatures are likely to benefit. (One potential candidate is the desert bighorn sheep, whose populations in interior California are dwindling as temperatures rise.)[12] No one is likely to make a life's work fighting for a rare variety of slug or dung beetle. This political reality weakens the concept of assisted migration as a way of preserving biodiversity—unless people start to study and battle for even the most humble creatures, much will be lost as habitats heat up.

Yet big, popular animals are often keystone species, with an outsized impact on their habitats, and could make a real difference in buffering ecosystems from climate change. Careful observation of hundreds of kills made by Yellowstone's reintroduced gray wolves provides evidence of this. Native scavengers—ravens, bald and golden eagles, magpies, coyotes, black bears, and grizzlies—all feed regularly at wolf kills. Before the wolf returned, the availability of carrion depended on the depth of snow and the length of the winter season. In winters of deep snow, elk struggled to find forage and often died of starvation. Carrion from these winter-killed elk was plentiful only late in the season or during severe winters, and over the past 50 years, Yellowstone winters have grown milder and shorter. In a wolfless park, that often meant starvation for scavengers.

Wolves now occupy most of the park and have become the main cause of elk mortality in Yellowstone. Chris Wilmers, an ecologist at University of California–Santa Cruz, has found that as they hunt throughout the winter, wolves leave rich pickings behind for scavengers. In late winter, often the hungriest time of year, they can increase the availability of carrion by more than 50 percent.[13] Although climate change may still mean less food for the coyote, the raven, and their ilk, wolves give scavengers more time to adapt.

The wolves' steady take can also buffer their prey from some impacts of warming. Wilmers has studied the interactions of wolves, weather, and prey in Yellowstone and at Isle Royale National Park, where the top dogs dine on moose. When wolves are few or absent, climate drives boom-and-bust cycles of large herbivores. Each time a population of moose or elk crashes due to a harsh winter or a summer drought, it loses genetic diversity and becomes more vulnerable to extirpation. Top predators prevent population booms among their prey, so that a harsh change in climate kills fewer animals.

By protecting carbon-storing plant life from out-of-control herbivore populations, predators may even have the power to slow the process of global warming. Perhaps the best example of a carbon-friendly predator is the sea otter, a species key to the health of kelp forests throughout much of its range along the Pacific coasts of North America and Russia. Otters prey on sea urchins, spiny invertebrates that feed on kelp. Where otters have disappeared, urchins boom, devouring every available scrap of kelp. The bleak seascape left behind is known as an urchin barren. Working with James Estes, the ecologist who first described the otter–urchin–kelp relationship, Wilmers has built a model of the otters' potential to help pull excess carbon dioxide out of the atmosphere.[14]

Hunted to near-extinction by nineteenth-century fur traders, sea otters survive in Alaska and on islands off the Russian coast. A few small, scattered populations also live off of British Columbia, Washington, Oregon, and California. Based on a scenario in which sea otters had been restored to the whole of their original range, Wilmers calculated how much carbon would be locked up in kelp forest. In his model, kelp forests would absorb about 40 percent of the carbon released to the atmosphere by humanity since the onset of the industrial revolution. That is a carbon sponge worth more than $700 million on the current European carbon market.

No terrestrial predator has an estimated impact on carbon balance as powerful as the sea otter's: plants of some sort will grow on most patches of earth, whether or not predators roam, but without the otter, huge swatches of seabed become, and remain, barren. Still, the wolf, the mountain lion, and the jaguar help increase absorption of carbon on their turf. In another thought experiment, Wilmers calculated that a global restoration of top predators would create a landscape capable of locking up 23 gigatons of the human-released carbon now loose in the atmosphere. The numbers may describe a goal achievable only in dreams, but there is an important take-home lesson nonetheless. "Burning fossil fuels has a serious impact," says Wilmers. "But so does getting rid of predators."

Carbon balance may be shifting fastest in the Arctic. Working outside of Barrow in the 1970s, ecologist Walter Oechel discovered that the tundra ecosystem there had been sucking carbon dioxide out of the air and locking it away in cold storage for 10,000 years, since the end of the Pleistocene. He figured this out by using radiocarbon techniques to date the peat lying beneath the dwarf shrubs, bright sedges, and thick mosses of the tundra.

A decade later, intrigued by the possible impacts of human-made global warming, Oechel returned to test the effects of rising temperatures on Arctic plant communities. He found that the terrestrial Arctic had reversed itself, switching from a carbon sink to a source.[15] The implications are staggering: the summer-thawed soil that supports the life of the tundra holds an estimated 60–190 billion metric tons of accumulated carbon, now slowly beginning to leak into the air. That dwarfs the amount released by human combustion of fossil fuels, the major driver of the annual three to four billion ton increase in atmospheric carbon over the past twenty-five years.

Tundra releases carbon as soil microbes break down organic debris. The microbes remain active at temperatures as low as minus 40 degrees Fahrenheit. Heat them up, and their respiration rates skyrocket, along with their release of CO_2. In recent years, as Arctic summer temperatures climb, tundra moss has given way to dwarf birch and willow, which grow taller in the warmer air. This creates a positive feedback loop that magnifies the impact of warming. In winter, the shrubs trap deep drifts of snow, which insulate the soil below. This keeps soil microbes warm enough to accelerate their digestion of organic litter throughout the winter, pumping out more CO_2 in the process.[16]

An even greater store of carbon—the frozen remains of the steppe vegetation that once fed woolly mammoth, wild horse, and long-horned bison—lies locked in permafrost, beneath the thin layer of biologically active soil that feeds the tundra. In parts of the far north, rising summer temperatures have begun to melt the ancient permafrost layer, which contains a supply of organic carbon 2.5 times greater than that of all the world's rainforests combined.[17] If current trends continue, tons of long-preserved mammoth fodder will be released into the modern atmosphere as carbon dioxide, accelerating the already dangerous rate of global warming.

In a remote corner of Siberia, a maverick ecologist is trying to stop the hemorrhage of Arctic carbon. Sergei Zimov believes he can use modern megafauna to revive the lost grassland that once fed the woolly rhinoceros and mammoth—and keep permafrost intact in the process. At the Northeast Scientific Station at Cherskii, in the Republic of Yakutia, he has begun a project he's optimistically dubbed Pleistocene Park. The goal is to introduce a diverse range of large mammals, to imitate as much as possible the fauna that lived near Cherskii during the Ice Age. Bring back large herbivores, he argues, and the landscape will grow more productive, while the underlying permafrost will be protected from the heat.

At first glance, the idea seems ridiculous—like Barrow, Cherskii is surrounded by a mosaic of soggy tundra and meltwater ponds, a bleak environment that seems incapable of supporting megafauna. But abundant bones of ancient mammoth, woolly rhino, bison, and horse show that this part of Siberia was long a part of the mammoth steppe—an arid, cold, but biologically productive prairie. As a student, Zimov was taught that the steppe vanished, morphing into tundra, due to warming temperatures and rising humidity at the end of the Ice Age. Yet indicators of prehistoric climate suggest that northern Siberia is no wetter now that it was during the Pleistocene. (Tundra soils stay wet only because underlying permafrost traps water at the surface. Rainfall around Cherskii is no higher than in some desert ecosystems far to the south.) Because the steppe faded away as the giants died out, Zimov suspects they were the architects of their own habitat. Their passing could not have been caused by climate change alone, he argues, since they had weathered many previous interglacial warm spells. The critical difference at the close of the Pleistocene was the arrival of people, armed with razor-sharp stone spearpoints.

According to Zimov, mammoths and other large herbivores kept woody shrubs from growing up, and by devouring grass they encouraged the continual resprouting of new stems. The ever-growing grass drew water from the earth and released it into the atmosphere, a process called evapotranspiration. With the loss of the megafauna, grazing dropped off drastically, and grass stems withered without being cropped. Throughout the steppe, evapotranspiration rates plummeted, and the ground became soggy, forming a perfect habitat for tundra mosses.[18] Few animals eat moss, and it acts as a literal wet blanket, slowing the growth of other plants and releasing little moisture to the air. Zimov and his colleagues have shown that adding nutrients to modern tundra vegetation boosts its productivity and evapotranspiration rates, drying out the soil below. Bringing horses or other large grazing animals onto a patch of tundra has the same effect: their heavy hooves break up the moss, their dung fertilizes the soil, grasses sprout and funnel moisture into the air, and the soil dries.

Part of Zimov's inspiration is a drive to create a sustainable way of life for people in northeastern Siberia. Yakutia's economy has always relied on reindeer. Prehistoric people there shadowed reindeer migrations, ambushing the animals at river crossings. Over time, they learned to tame the snow deer, using them as mounts and beasts of burden. But these animals always teetered on the brink of wildness and might escape at any time to run free with wild herds. It was not until the twentieth century, when reindeer farming came under the control of the Soviet government, that the creatures became truly domesticated.[19] Under a government-subsidized program of intensive breeding, domestic herds grew to a peak of 2.4 million animals in 1970. When crisis struck the Russian economy in the 1990s, domestic reindeer numbers plummeted and had dropped to 1.2 million in 2002. The people of Cherskii fell on hard times, and many have had to leave to find work elsewhere.

To raise initial funding for the Pleistocene Park experiment, Zimov approached the Yakutian government and pitched the idea that in a world of changing environments, a heavy reliance on reindeer herding no longer made sense. If he could develop an ecosystem that harbored bison, horse, and other large creatures, the megafauna would in turn sustain local people. Zimov and his family are utterly committed to this quest—they live in Cherskii year-round, rather than escaping to a city during the dark winters, as many Arctic researchers do.

The project began in 1989, when Zimov introduced a small herd of Yakutian horses, the closest descendants of the local Pleistocene horse, to a paddock at the science station. The success of horses in the region is, for Zimov, confirmation of his theory that climate change alone did not eliminate the Pleistocene megafauna. Today, the biomass of free-roaming horses in Yakutia has outstripped that of reindeer. Though the horses are classified as a domestic breed, they seem well adapted to survive in the wild. As Zimov had hoped, they transformed their pasture, devouring sparse grass and fertilizing its regrowth with manure. At first the horses were fed hay, but over time enough grass grew up to sustain them. Years after they were released to roam a large fenced enclosure, their original paddock remains a fertile grassland. Those first horses and their descendants now live without being fed by people, grazing in patches of grass they helped to create, scattered across the expanse of Pleistocene Park.

Zimov has not yet documented a similar trend outside the original small horse paddocks, however, and many of his colleagues remain skeptical. So there was excitement in Cherskii when researchers working on a separate study, far across the Arctic, confirmed his ideas.

In West Greenland, as elsewhere in the far north, warming boosts the growth of dwarf birch and willow and leads to a decline in grasses and sedges—the beginning of the carbon-releasing feedback loop studied by Oechel and others in Alaska. That, at least, is the response when an area is fenced to keep out musk oxen and caribou. Where the big grazers roam, results are dramatically different: the vegetation does not change at all. The animals chew up any shrubs that sprout in response to climate change, explains Eric Post of Pennsylvania State University, who in 2002 set up an experiment to compare the response of tundra plants to artificial warming, with and without the presence of large herbivores.[20]

Post expected caribou to control the ecosystem response to warming. To his surprise, musk oxen—at the start of his experiment much scarcer than caribou—did all the work. (Several years later, caribou have all but disappeared from the area, a sad testimony to the species' global decline.) The caribou is a restless, picky eater, constantly on the move as it plucks the choicest bits of grass, flowers, and lichen. The musk ox, by contrast, devours large amounts of whatever lies before it. Many Arctic plants are indigestible and simply come out the other end. But the strategy works for the musk ox and, it seems, for its habitat.

Where musk oxen graze, the vegetation is dominated by a highly productive, diverse range of grass and sedge, enriched with nitrogen left

Fig. 29 Musk oxen, the closest living stand-in for the extinct woolly mammoth, shape Arctic vegetation. Some researchers believe musk oxen and other large herbivores can help prevent the release of huge stores of carbon beginning to melt out of permafrost. (Photo courtesy of U.S. Fish and Wildlife Service.)

behind in animal waste. Fence the musk oxen out, and woody shrubs take over, shading out more nutritious forage plants.

Post describes the musk ox as "a furry jeep with horns." In ecological terms, he sees them as the best surviving stand-in for the vanished woolly mammoth. They trample and break shrubs, using them not only as food but as combs and back-scratchers. Woolly mammoths likely behaved in similar ways, but because they were much larger and heavier, a single animal would have had far more impact. Based on his work in Greenland, Post agrees with Zimov that large herbivores could play an important role in mitigating the impacts of modern climate change, especially in the far north.

Much of Pleistocene Park remains soggy tundra, but the land is now populated with reindeer and moose, as well as a thriving horse herd. Zimov hopes to import bison from Canada or Alaska, as well as musk oxen and the Siberian goat, which once lived in Yakutia's lowlands. (Native goats survive only in mountains, where there are no people to disturb them.) To succeed, says Zimov, the park needs a diversity of animals that feed on every kind of local plant—the rest will be a matter of time. Someday, when the big herbivores have flourished and multiplied, he dreams of bringing rare Siberian tigers in to fill the vital role of top predator. These Pleistocene survivors have vanished from much of their original range, but a few hundred hang on, sometimes allegedly killing dogs or livestock on the fringes of Vladivostok. Wolves will also help fill the predator's niche, though there will be no need to reintroduce them. In the early days of the experiment, wolf predation knocked the horse population back so hard that Zimov's colleague, Sergei Davidov, spent a good portion of his time wolf hunting.

Zimov is an enthusiastic supporter of Pleistocene rewilding in America—or most anywhere, if it can be made politically viable. In his view, even the remotest landscapes in today's human-dominated world are not wild, but ecologically dead. That very much applies to Cherskii's sodden tundra. (It is a family joke that he came up with the Pleistocene Park hypothesis while hiking ankle deep in muck, wishing he lived in better walking terrain.) For Zimov, the dream of resurrecting the biological intensity of the Pleistocene justifies taking risks, mixing megafauna from different parts of the planet.

The vast sweep of the north has always made a good backdrop for wild dreams. There are some who would push beyond Zimov's grand ambition of restoring Siberian grassland. They believe they can resurrect the long-dead woolly mammoth, bringing the original ruler of the Ice Age steppe back to its frigid stomping grounds.

DEAD
BEASTS
WALKING

THE MAMMOTH DECIPHERED

In late 2008, Webb Miller and Stephan Schuster, geneticists at Pennsylvania State University, announced a breakthrough in the study of ancient DNA: they had sequenced most of the woolly mammoth genome.[1] Seemingly intact bodies of Pleistocene megafauna—including mammoths and extinct bison—sometimes work their way to the surface of Arctic permafrost. Though these ice mummies may look perfect, the molecular spirals of DNA that hold the blueprint for the living animal unravel soon after death, leaving only scattered fragments. Until Miller and Schuster developed a technique to decontaminate and extract long strands of DNA from preserved hair, few in their field believed it possible to reconstruct the entire genetic code of a long-extinct beast.

The researchers used a draft map of the African elephant genome to help trim out erroneous bits of DNA that had crept into their samples from bacteria and fungi. Now, says Schuster, scientists find themselves in the strange position of knowing more about the genetic makeup of the extinct woolly mammoth than they do about its living cousins, the African and Asian elephants. In addition to the near-complete nuclear genome, the entire sequences of mitochondrial DNA (mtDNA) from eighteen individual mammoths have been deciphered. (Mitochondria are organelles that power living cells, burning oxygen to release energy. Each cell contains hundreds, sometimes thousands, of mitochondria; each mitochondrion carries a brief stretch of genetic code. Because the sequence is short and abundant, mtDNA is far easier to reconstruct from ancient samples than is the longer, scarcer nuclear genome.)

Mitochondrial DNA reveals a striking shift in mammoth genetics that took place about 45,000 years bp, long before the species began the steep decline in numbers that would mark the beginning of its end. Until that distant moment, two genetically distinct groups of mammoth—dubbed clade I and clade II—had coexisted in Beringia. Then clade II vanished, for reasons unknown, and the inheritors of clade I genes spread west out of Alaska and the Yukon, populating the whole of the mammoth steppe.[2] The mysterious demise of clade II meant that millennia later, when drastic change struck the mammoth population, it had already lost much of its genetic diversity.

At the moment of crisis, when the glaciers retreated and a wave of Arctic-adapted people flooded north in their wake, the mammoth's fate may already have been sealed by this genetic loss. Today, conservation biologists strive to maintain genetic diversity among endangered species. A population whose members carry a variety of forms of important genes has untapped flexibility, and therefore the hope of finding new ways to respond to environmental change.

This window into mammoth genetics has inspired Schuster and Miller to further explore what they call the "biology of extinction." From studies of the woolly rhinoceros and its rare living cousins, the lost moa of New Zealand, and Australia's extinct thylacine, the researchers see a clear pattern emerging. As a species draws closer to extinction, its genetic diversity plummets. This means that in living populations of rare creatures, loss of genetic diversity is a critical warning sign.

Schuster and Miller are applying the lessons of the mammoth's demise to the Tasmanian devil, the largest surviving marsupial carnivore. Solitary through most of their lives, devils come together to share finds of carrion and to mate, a fast, passionate process during which they bite and slash at each other. Adults are about the size of a Welsh Corgi dog but with fierce-looking, pointed snouts. The species once roamed throughout Australia, but the last mainland populations died out more than 300 years ago.[3] (A giant form of devil disappeared long ago, with the rest of the continent's Ice Age megafauna.)

Devils survive in the wild only on the island state of Tasmania. Populations there are rapidly dying off, victims of a virulent transmissible cancer, devil facial tumor disease (DFTD). The cancer, first documented in 1996, originated in one hapless beast somewhere in eastern

Tasmania. Passing from one animal to the next during violent mating bouts, it establishes itself in the facial wounds that are an inherent part of sex among devils.[4] DFTD is a classic example of the dangers of genetic uniformity: members of the population are all so alike that their immune systems do not recognize the cancer as foreign, even though it came from another animal. (Surviving devils are so genetically similar they might all successfully donate organs to one another.) The disease has rapidly brought the Tasmanian devil to the brink of extinction.

Schuster and Miller sequenced mtDNA from modern devils scattered across the species' range and from museum specimens up to a century old. The devils, they discovered, had lost much of their genetic diversity decades ago. The pair is now creating a gene-based protocol for breeding devils in captive populations, designed to mate animals that are as different from each other as possible to maximize the genetic diversity of the offspring. These purposefully diversified animals may represent the species' best hope of developing an immune response to the contagious tumor, and of long-term survival. This kind of insight—a lesson from the Pleistocene that can help to keep modern creatures alive—is Schuster's driving motivation.

For some, the mapping of the mammoth genome summoned visions of a very different sort of resurrection. The idea of bringing the mammoth itself back to life has been around for years, since the early days of bioengineering. In the 1980s, Russian scientists tried to clone cells from Dima, a frozen baby mammoth discovered by Siberian miners in 1977. They hoped to extract a nucleus holding intact chromosomes from one of Dima's cells, transplant it into an elephant egg emptied of its own nucleus, zap the cell into action with electricity, and transplant it into the womb of an elephant surrogate mother. But months of mining Dima's best-preserved tissues—muscle, subcutaneous fat, blood—failed to produce a single viable nucleus.[5]

In the late 1980s, Kazufumi Goto, a Japanese expert on *in vitro* fertilization, managed to engineer a live calf, the fruit of a cow's egg fertilized with a dead sperm cell. Goto reasoned that if frozen, dead bull sperm could accomplish this feat, a sperm cell from a frozen mammoth might be able to repeat the miracle with an elephant egg. Carried to term by a surrogate elephant mother, the product of such a union would be a half-elephant, half-mammoth hybrid. He planned to produce only female calves, harvest their eggs, and cross them *in vitro* with more dead mammoth sperm. After a few rounds of back-crossing, he would have a creature that was almost pure mammoth.

Goto's work inspired the formation of the Mammoth Creation Project, a group of scientists and businessmen that has traveled to Siberia in search of a well-preserved male mammoth.[6] On one of their journeys they explored the rich deposits of Pleistocene fossils near Sergei Zimov's science station in Cherskii. One flaw in their plan, as Zimov points out, is that every near-intact mammoth carcass has been found by accident, by miners or reindeer herders who happened to be in the right place at the right time. A directed search for a quick-frozen mammoth—the only kind of specimen that could provide usable sperm—is worse than hunting a needle in a haystack.

Another major glitch in Goto's vision is that a dead mammoth will freeze and thaw repeatedly before it is incorporated into the permafrost, a process that breaks down chromosomes and other complex molecules. Arctic microbes remain active at subzero temperatures, degrading tissues even after they have frozen. So it's no surprise that no one has ever found a cell containing an intact nucleus, even in the best-preserved mammoth carcass.

Still, the sequencing of a near-complete mammoth genome opens other avenues for creating a live mammoth. Geneticists could engineer a mammoth from scratch, synthesizing the genome and forming it into chromosomes, enclosing the whole package in an artificial nucleus within a modified elephant egg. Jolted into cell division by electric shock, this man-made embryo could be implanted in an elephant surrogate mother. Numerous technical obstacles, however, make it unlikely that this scenario will be carried out any time soon.[7]

Another possibility would be to extract the chromosomes of a modern elephant and modify them, bit by bit, to match the DNA sequence found in the mammoth. Schuster estimates that it would take 400,000 alterations to tailor an Asian elephant genome into a replica of the mammoth's. George Church, a genomics pioneer based at Harvard and the Massachusetts Institute of Technology, envisions a much simpler endeavor. "You can be very fussy and insist on getting the genome exactly right," he explains, "or you can go for something that has the main visible characteristics: the hair, the size, the tusk shape." Getting something that *looks* like a mammoth might entail a relatively small number of gene changes. Church's lab is already altering mouse genomes bit by bit, remaking them 100,000 DNA base pairs at a time. No one has actually turned a mouse into some other creature yet, but the Church group is well on its way to morphing a mouse into a naked mole rat, a fascinating rodent that lives ten times longer than a mouse, in elaborately structured social hierarchies ruled by queens.

The prospect of mammothified elephants walking the earth makes Schuster uneasy, but he believes it will come to pass. "As we speak today, it is not possible to resurrect a mammoth. It is nothing that any of us is pursuing," he says. But reproductive technology is improving so quickly that creating some version of the mammoth may well become possible soon—in part because of intense efforts to assist the breeding of endangered Asian elephants in captivity. Because elephant reproduction is slow and complex, and handling male elephants in musth is so dangerous, this necessarily becomes a high-tech process.

A mammoth, by nature a deeply social creature, might not thank us for bringing it to life alone, on a planet nearly devoid of its native steppe habitat. Perhaps more worrisome is the prospect of faux mammoths, or other extinct creatures, which might too easily be accepted as replacements for the genuine wild beasts. Looking like the real thing does not mean an engineered animal will act, or survive, like the real thing in the wild.

In an elegant experiment, researchers recently revealed a critical, but invisible, difference between the mammoth and the modern elephant.[8] Using ancient mammoth DNA, they reconstructed the creature's hemoglobin, the blood protein that picks up oxygen in the lungs and offloads it in the tissues. Like other Arctic animals, the mammoth kept its feet much colder than its core body temperature, to avoid losing body heat to the frozen earth. This was possible because they had evolved an altered form of hemoglobin that readily releases oxygen under cold conditions. The hemoglobin produced by modern Asian elephants, however, cannot release oxygen at such temperatures. The fact of mammoth hemoglobin breathing in twenty-first-century air is thrilling. Yet the functional differences between the mammoth and its living cousin underscore potential problems with tailoring a mammoth-like elephant for conservation purposes.

Such an animal might fit in to its ecosystem no better than Frankenstein's monster did his local village.

Enthusiasts seem unperturbed by the ethical problems associated with mammoth resurrection and oddly uninterested in finding a sound justification for the process. To Church, raising the mammoth is simply an engineering goal, with no fundamental obstacles in the way except the will to forge ahead. "To prove we can bring a species back from extinction I think is quite valuable," he says. "We should be developing technology now to bring them back, because we may need it on short notice." He suggests it would be possible to create both males and females, which would breed and found new populations, reviving lost biodiversity. Yet he seems quite willing to settle for a creature that merely looks like a mammoth.

Since the first scientists toyed with the notion, bringing the mammoth back to life has been more about drama than ecology. The dream of raising the lost Pleistocene megafauna has been around for decades, long before anyone even imagined genetically engineering a living being. But the shifting shape of that fantasy reveals more about people than about any other kind of megafauna.

THE STRANGE TALE OF THE NAZI COW

Mammoth hunters like Goto are not the first to dream of resurrecting an extinct giant, or to put years of work into such a project. That distinction goes to the brothers Lutz and Heinz Heck, zoologists and enthusiastic members of the Nazi party in the days of the Third Reich. They claimed to have revived not one, but two, long-extinct species.

Europe lost some impressive megafauna at the end of the Ice Age, including the woolly rhino, the mammoth, and the giant deer with its humped back and massive antlers. Other great beasts lived on to inhabit human legends and even transform human societies. Foremost among these was the aurochs, a massive wild ox that gave rise to domestic cattle, and the tarpan, Europe's native horse. Though they were carefully protected by royal order, the last herd of aurochsen died out in the year 1627, at the now-vanished forest of Jaktorow in Poland.[1] Fleeing from hunters, the last wild tarpan mare fell to her death in a crevasse in Ukraine in 1880.[2] Both species had dwindled from centuries of hunting by an ever-increasing human population. Even more deadly was the impact of domesticated livestock, including, ironically, their own descendants. People and their herds pushed wild megafauna out of prime grazing lands, leaving the majestic aurochs to eke out a living in marginal habitat. Evidence of this shift has recently been uncovered in fossil bones and teeth. Isotope analysis shows that aurochs and domestic cattle of the same era fed on different kinds of plants. Patterns of tooth wear from aurochs in the last few centuries of their existence suggest they were forced out of the open grasslands they preferred and had to forage on less nutritious shrubbery at forest edges.[3]

The Heck brothers shared an obsession with these great lost creatures and a belief that they could be brought back to life. They assumed that the entire biological legacy of the wild aurochs and tarpan lived on in their domesticated descendants, divvied up among different breeds. By crossing various breeds of domestic cattle and horse, they reasoned, they could produce offspring that would carry the whole, reassembled inheritance of the extinct creatures. Both brothers began breeding experiments with domestic cattle in the 1920s—Heinz working at the Munich Zoo, Lutz at the Berlin Zoo. By 1933, when Hitler came to power, Heinz claimed he had produced two genuine aurochs bulls.[4]

Of course he was wrong: what he had was a new domestic breed, now known as Heck cattle, whose coat color resembled that of the lost aurochs (russet in cows, black in bulls, with a pale line along the backbone). Striking images of aurochsen were left behind by Pleistocene cave painters at Lascaux in France and Altamira in Spain. Tens of thousands of years later, these artworks still convey the intimidating size and spirit of the real beast. A wild aurochs bull would tower over a herd of Heck cattle (old bones show that the largest aurochsen stood as tall as 6.5 feet

Fig. 30 This Pleistocene cave painting from Lascaux, France, shows wild aurochsen, horses, and reindeer. (Photo from Wikimedia Commons, http://en.wikipedia.org/wiki/File:Lascaux_painting.jpg.)

at the shoulder). They were longer of leg, bigger of brain, more graceful and fearless than their domesticated brethren. Thousands of years of selection for docile cows that produce an excess of meat and milk couldn't really be undone—certainly not by a dozen years of cattle breeding. But the Hecks made their claim boldly, and it served them well under the Nazi regime.

Along with their passion for producing and safeguarding a "pure" race of Aryan people, Hitler and his lieutenants yearned for a return to an imagined, pristine state of nature. In the Nibelungenlied, an epic poem that obsessed Hitler, the protagonist hunted aurochs and tarpan and wore a helmet made of aurochs horn. The Nazis greeted with enthusiasm the Hecks' announcement that they had recreated the aurochs, and Lutz in particular cultivated connections with the elite. Goering and Goebbels became his houseguests and hunting companions.

For officials high in the Nazi hierarchy, the mission of purging of people they deemed inferior from society paralleled the resurrection of Europe's native megafauna. Lutz Heck's colleague and friend Eugene Fischer was among the scientists who bought into this ideology and used the goal of racial purity to justify forced sterilization and mass murder of human beings. Interbreeding, Fischer said, was degrading humans in the same way it had already diminished "beautiful, good and heroic" wildlife.[5]

In addition to raising the aurochs and the tarpan, the Heck brothers wanted to rescue the European bison, which teetered on the brink of extinction when the war began. Only a handful of bison survived at zoos and reserves scattered across Europe, and careful work was ongoing to breed them in captivity. One of the leaders of the bison recovery effort was Jan Zabinski, director of the Warsaw Zoo. In 1929, a few of Zabinski's bison were reintroduced to the wild at the Bialowieza Primeval Forest, a reserve considered a Polish national treasure.

In one of his strangest experiments, begun in 1927, Lutz Heck crossed the highly endangered European bison with Canadian bison. His goal was to transfer what he called "the immense reproductive energy of the present-day American bison" to the European breed.[6] This was an odd tactic, given that most mainstream conservationists then, as now, treasured and protected native wild creatures, not exotic ones. Heck emphasized that once he had introduced the required dash of vitality, he would be careful to select against any American-looking animals in his herd. Although his breeding experiments were based more on opinion and eccentricity than on science, it seems he understood biology well enough to sense the importance of diversity. It would not have been politic to say so when Hitler came to power, but a pure breed is often, in the long run, a doomed breed. The European bison, today still struggling to hold on in a few highly inbred, disease-prone populations, is living proof.[7]

The war gave Lutz Heck the opportunity to loot rare animals from occupied countries, seizing every beast that could contribute to his project, funded by Goering, of recasting nature according to a Nazi ideal. Soon after Germany invaded Poland in 1939, Heck appeared at the Warsaw Zoo, which had already lost many animals during weeks of bombing. He took every beast he considered valuable, and was particularly happy to get his hands on the Przewalski's horses (members of a near-extinct variety of Eurasian wild horse), which he sent to Vienna. He also took the zoo's European bison.

One sunny December day, Heck returned to the Warsaw Zoo with a group of Gestapo members and fellow hunters. They spent hours there, drinking vodka and shooting caged zoo animals for sport, wiping out the creatures that had been in Zabinski's care.[8]

By 1942, Lutz Heck's pseudo-aurochsen had become so numerous that he decided to have a herd released into the wild at Bialowieza. He also set loose thirty of the bison he had stolen from European zoos, along with some of his brother's back-bred tarpans. He envisioned Bialowieza, with its massive, moss-draped oaks, as an Aryan haven where Goering and Hitler would ride to the hunt.[9] That never came to pass, and in the hard times that followed the war's end, all the Heck cattle at Bialowieza were shot for food. (Groups of Heck cattle survived the war at the Munich Zoo and elsewhere.)

History is full of ironies. Jan Zabinski, his wife and son remained at the empty Warsaw Zoo through most of the war, using its grounds to hide Jews whom Jan helped escape from the ghetto. When the war ended, he continued to work with the International Society for the Preservation of European Bison. The Allied bombing of Germany killed captive herds of bison there, leaving those in Bialowieza—the animals Heck had planned to offer up to the Führer, along with his home-bred aurochsen— as the species' best hope for survival.

The story of the Heck brothers' attempt to resurrect the aurochs and the tarpan might seem like no more than an oddly twisted footnote to the millennial panorama of changing megafauna. But in another bit of irony, Heck cattle have been adopted by European conservationists who believe the animals can restore long-lost ecosystems. One such activist, Derek Gow, imported the first Heck cattle to Britain in the spring of 2009, hoping to breed them and supply animals to nature reserves. The *London Times* covered the event. "How a Nazi Experiment Brought Extinct Aurochs to Devon," read the headline, again blurring the line between man-made megafauna and the genuine, long-lost beasts.[10]

The megafauna we tamed—the smaller, meeker descendants of aurochs and tarpan, wild boar, wolf, wild goats, and sheep—made humans the world-dominating creatures we are today. Soon after the Ice Age ended, people began, for the first time, to cultivate grains and to herd sheep, goats, and cattle. The oldest known traces of livestock domestication come from western Asia and are about 10,000 years old.[11] The driving forces behind the transition from foraging to agriculture remain hotly debated. This radical shift in survival strategy coincided with a global warming trend at the beginning of the Holocene, which increased the abundance of edible, storable plant foods.[12] As their numbers grew in the mild new climate, people began to domesticate both plants and animals and to shift from a nomadic life to a settled one.

Whatever factors drove the shift to farming, it happened among peoples widely scattered around the globe at the same moment, around 10,000 years bp. Genetic studies that compare ancient DNA of fossil aurochs and modern cattle show that aurochs domestication occurred at least three separate times, in Eurasia, Africa, and the Fertile Crescent of the Near East.[13] DNA evidence of cattle domestication is also found in people—only the descendants of cultures with a long history of keeping dairy animals continue to produce the enzyme lactase beyond early childhood, allowing them to efficiently digest milk throughout their lives. People in Africa and Europe achieved the trait of lactase persistence through different genetic mutations, a striking example of convergent evolution in humans and proof that domesticated megafauna have shaped us just as we have shaped them.[14] The evolution of farming created reliable stocks of food that could sustain more people, and the increase in hungry mouths to feed drove the expansion of agriculture. In a self-perpetuating cycle, sheep, goats, cattle, and horses fueled a global expansion of human populations.

In much of Europe, the triumph of agriculture was so complete that no pristine places remain as reference points to hint at the shape of undisturbed nature. When the first European conservation organizations started up at the turn of the twentieth century, they created nature reserves that were often fenced off and left alone. Devoid of megafauna, these areas became overgrown with weeds, leading to a decline in the diversity of birds, small mammals, and insects. To avoid this, some managers began to mow the grass regularly, and in general treated their reserves like nineteenth-century farms, sometimes with and sometimes without livestock.

Decades later, most European conservationists agree that megafauna are essential to healthy ecosystems. By grazing and browsing, they maintain a diverse plant community, while their waste fertilizes the soil. Many reserves—including England's renowned New Forest—have long hosted cattle and horses along with wild deer. Recently, a cadre of Dutch researchers have pioneered a new approach, a European form of rewilding. They propose using hardy breeds of cattle and horse as stand-ins for the lost aurochs and tarpan. But in a radical departure from tradition, the herds are not managed, allowing their numbers to multiply beyond what has traditionally been viewed as the carrying capacity of the land.[15] The flagship example of this approach is the Oostvaardersplassen, 5,600 hectares (21.6 square miles) of reclaimed Netherlands marsh rich in bird life, where the dominant megafauna are elk, Heck cattle, and the Konik horse, a Polish breed developed to resemble the tarpan.

The Oostvaardersplassen, like most European reserves, lacks large predators. Wolves, the only viable candidate for this niche, were extirpated from much of the continent in the nineteenth century. Wolves preyed primarily on ibex and elk until those species vanished, their habitat co-opted by humans and their livestock. When the top dogs turned their attention to domestic sheep, people responded with rifles and strychnine.[16] The wolf survived only in a few wild havens in Russia and eastern Europe. They are now slowly returning to western Europe, and small populations live in rural Italy, France, Switzerland, and Finland, a development greeted with alarm by many farmers. No wilderness area

Fig. 31   A Heck bull surrounded by Konik horses at the Oostvaardersplassen, a reserve in the Netherlands where domestic livestock are being used as ecologic substitutes for extinct aurochs and native horse. (Photo from Wikimedia Commons, http://en.wikipedia.org/wiki/File:GroteGrazers.jpg.)

in Europe is large enough to hold a healthy population of wolves, and intentional reintroductions remain politically impossible.

The only control on large herbivore populations at Oostvaardersplassen are rangers operating with what some have called "the eye of the wolf." They survey the herds and, if an animal looks sick, weak, or injured, call in a veterinarian who decides if the beast has a chance to recover. If not, the animal is culled.

The saga of the Yellowstone wolves gives the lie to the idea that rangers following humane guidelines could mimic the impact of a wild wolf pack. Critics point out that all the grassland at Oostvaardersplassen is closely cropped, and despite fertile soils, no new trees grow up because hungry horses strip saplings of bark.[17] More than twenty years after cattle and horses were set free there, the reserve confirms that big herbivores boost biodiversity only when their numbers are controlled by effective predators. The scene advocates cherish as one of pristine nature may more closely resemble a depleted Yellowstone aspen stand in the years before wolves returned.

In Europe as everywhere else, conservation is entangled with human quirks, with unrealistic visions of life in the wild, and with the painful memories of history. Joep van de Vlassaker, a consultant involved in rewilding projects across Europe and a supporter of the Oostvaardersplassen approach, has fought a losing battle to bring Heck cattle back to Bialowieza—a proposal Polish patriots, who well remember the breed's history, will not tolerate. The Poles want to keep their most precious reserve as habitat for the highly endangered European bison, and evidence that aurochs and bison coexisted for millennia has so far failed to change their minds about bringing cattle back into the mix.

The origin of Heck cattle, van de Vlassaker feels, is something best forgotten. He acknowledges that any hardy breed of cattle could stand in for the aurochs—some reserves have successfully used shaggy Highland cattle, for instance. Heck cattle come with heavy political baggage and have difficult temperaments (one breeder describes them as "hideously fast and aggressive"),[18] but van de Vlassaker prefers them because they look like aurochsen. "It helps to give a clear message to the public why you have cattle in nature reserves," he explains.

If the idea of a conservation-oriented cow is difficult for some Europeans to grasp, the notion may strike American environmentalists as absurd. For more than a century, vast swathes of arid land in the western United States were managed for the benefit of cattle ranchers, and the results were often devastating. Thirsty cattle trampled and overgrazed narrow strips of riparian vegetation, collapsing stream banks and wiping out forage. By the early 1990s, cattle impacts were depleting wild populations of everything from trout to pronghorn antelope and elk. A report funded by the U.S. Environmental Protection Agency found that western riparian areas were in the worst condition in history.[19] A movement to completely remove cattle from federal rangelands gained momentum—summed up on a popular bumper sticker that read "Cattle-Free by '93." Long-time ranchers, who depend on public lands to sustain their herds, responded with outrage.

The western range has always supported large herds of hoofed animals. The native horse and camel may have been gone for 13,000 years, but the great herds of bison that supported Plains Indian society had been hunted out only a century before environmental activists began working to evict domestic cattle. Tens of millions of bison inhabited the Great Plains before the 1800s; from 1830 to 1880, a campaign of deliberate slaughter reduced their numbers to a few thousand. Much of the prairie habitat that had been home to the bison was plowed up by farmers at the same time.

As conservationists began working to protect and expand the few remnant scraps of native prairie, they experimented with bringing bison back to the landscape. A classic study of bison ecology at Konza Prairie in Kansas revealed the critical ways in which bison shape their habitats.[20] Their nitrogen-rich excrement boosts plant growth, and their grazing patterns create a diversity of plant species and growth stages, building niches needed by prairie birds. The lesser prairie chicken, for example, requires tall grass to conceal its nest, as well as open or short-grass areas for courtship displays. Other birds, like the western meadowlark, do best on patchy grasslands, with a mosaic of short and tall grasses mixed with wildflowers.[21] This kind of pattern results from a combination of periodic wildfire and grazing. Bison enrich the prairie—often through behaviors they share with the much-maligned domestic cow. Yet wild bison and the keepers of domestic cattle often clash.

Yellowstone's bison herd—the last free-roaming population in North America—embodies this conflict. The herd came close to extinction in the late 1800s, but by the mid-2000s it had rebounded to about 4,000 animals. As the population grows, bison migrate beyond park boundaries, where they are shot by agents of the Montana Department of Livestock. Many of the Yellowstone bison have been exposed to brucellosis, a bacterial infection that can cause pregnant cattle or bison to abort their calves. After decades of vaccinating and culling cattle, domestic herds in the United States are brucellosis free. Yellowstone bison, however, still carry the infection that a few animals picked up from free-range cattle early in the twentieth century.[22] Every winter, as the bison head out of the park in search of forage, hundreds are slaughtered to preclude any chance of their passing brucellosis to domestic cattle. In the winter of 2008, a third of the Yellowstone bison herd was shot.

Native bison clearly play a pivotal role in healthy prairie ecosystems. Still, despite the similarities between bison and cattle, environmentalists have continued to see the introduced cow as an agent of destruction. It is difficult to test this assumption, because the two grazers lead very different lives. Cattle graze the range during the spring and summer and in the fall are gathered up into paddocks where they spend the winter, feeding on hay. Bison, by contrast, live on the range year-round without supplemental feed. The first direct comparison of bison and cattle impacts was a ten-year study on Konza Prairie, which tracked the effects of grazing by small groups of both herbivores, managed in identical ways. The researchers found that both grazers increased the biodiversity of native prairie plants. The way each herbivore is managed, it turns out, plays a greater role in determining its influence than any innate differences between the species.[23]

The cow has been a plague on the West not because of its nature but because of the way people have traditionally kept it. In the United States, from the time barbed wire was invented in the 1870s, ranchers have fenced off pastures and let their cattle feed in the same area for most of the year. This is called the "Columbus school" of ranching: turn the cows out in May and then "discover" them in October.[24] This strategy is a recipe for overgrazing, especially along streams, where cattle linger. Progressive ranchers are now finding ways to keep their cattle on the move, to mimic the behavior of bison. Wild grazers roam from one patch of grass to another, giving the soil time to absorb the nutrients the animals have left behind and the plants time to regrow. During the Pleistocene, herds were free to follow the shifting flush of fresh grass, and saber-toothed cats and dire wolves helped keep herbivores alert and on the move. Now, people must intentionally shift their cattle from one fenced pasture to another or herd free-roaming animals with the help of dogs. These grazing systems demand more effort and attention than the Columbus approach. For a growing number of ranchers, however, the extra effort is proving worthwhile, improving the health of grasslands, cattle, and wildlife.[25]

Ranchers and environmentalists have begun to move beyond mutual fear and loathing, finding ways to work together. The Nature Conservancy, which forty years ago saw removing cattle as key to the revival of western grasslands, has started to use carefully managed cows as substitutes for native herbivores on some of its reserves. The Conservancy learned through trial and error that simply removing grazers would not restore wild habitats. In the 1970s, they fenced off rare populations of the near-extinct ladies' tresses orchid in an Arizona desert marsh, to protect them from stomping, chomping cows. Outcompeted by spikerush, a water-loving plant that flourished in the absence of large herbivores, the orchids dwindled from several hundred plants to a single flower.[26]

As human populations grow in the arid West, the lands kept open by ranching are increasingly being sold off and built up into suburban ranchettes. The pressure to sell is intense: cattle prices have been low for years, while the costs of running a ranch continue to climb. Most land-owners who hold permits to graze their stock on federal land must support their ranching habit with some kind of outside work.[27]

Yet a well-run ranch holds far more potential for biodiversity than a residential zone does, and some, like the 502-square-mile Diamond A Ranch in Arizona, are wildlife hotspots. Lying at a rich biological intersection where creatures from the Rocky Mountains, Great Plains, Chihuahuan Desert, and the Sierra Madre mingle, the Diamond A hosts 75 species of mammals, 52 different kinds of reptiles and amphibians, more than 700 plant species, and an array of birds—a third of all known U.S. bird species have been reported there.[28] The ranch is also the site of a shared grassbank, a tract of protected range where neighbors can bring their cattle to graze, allowing them to rest their own pastures in times of drought. In return for grazing privileges on the Diamond A, ranchers sign onto a conservation easement that precludes them from selling their own land for subdivision or development. This creative concept was brought to the range by the Malpai Borderlands Group, a confederation of ranchers who live along the Continental Divide where Arizona and New Mexico meet the Mexican states of Sonora and Chihuahua. The members of the Malpai Borderlands Group raise cattle amid one of the widest stretches of open space left in the United States, though development is creeping up along its edges. Searching for a way to keep the land open and their way of life intact, the ranchers banded together in the mid-1990s and invited scientists and environmental groups, including the Nature Conservancy, to help them manage for healthy wildlife as well as cattle.

The formation of the Malpai Borderlands Group was motivated by the belief that the range would continue to deteriorate in the absence of fire.

For more than a century, federal agencies and individual ranchers had snuffed out every spark of wildfire. Native grasses are fire-adapted, and the absence of flames has encouraged an invasion of mesquite, juniper, and sagebrush in once wide-open grasslands along the U.S.–Mexico border.[29] These woody plants are native to the region, but they have boomed in the decades since European settlement. Ranchers are deeply concerned about the spread of these shrubs, which leaves less forage for cattle. Climate is a major driver of the shift in vegetation—intensified winter rains, coupled with dry summers, favor shrubs over grasses. Growth rings of pine trees indicate that recent levels of winter rain are the highest in 2,000 years.[30] Still, there is good evidence that wildfire and large grazers—which once included abundant bison—played a part in keeping the range open before the mass livestock boom that hit the Southwest at the end of the nineteenth century.[31] Cattle, sheep, and their fire-averse human keepers have helped bring about the brush invasion.

Livestock grazing in the Southwest mushroomed in the 1870s and 1880s, after the region's grasslands were declared public domain, free to all takers. Cattle and sheep populations exploded: the count in New Mexico alone rose from 41,000 cattle and 619,000 sheep in 1870 to around 800,000 cattle and 5 million sheep in 1885. The resulting level of grazing far surpassed the impacts of native bison. Overgrazing, followed by drought, led to the demise of up to 75 percent of the herds in the early 1890s. At that point large areas that had been covered in perennial grasses were transformed to bare ground.

In the aftermath of this disaster, adherents of the new discipline of range science began to promote fencing the landscape and limiting the number of cattle permitted on each allotment. The range was divided to prevent another destructive grazing free-for-all, but the fences also broke up habitat for pronghorn, mule deer, and bighorn sheep, leading to a decline in their numbers. Ranchers' forced investment in expensive barriers built with flammable wooden posts made them eager to quash all wildfires, thus further degrading the health of the range. Fences also led to the development of the Columbus school of ranching, which would create its own set of overgrazing problems.

Banding together has allowed the Malpai ranchers to steward the broader landscape, rather than focusing on their individual plots of land. Cooperating with federal agencies, they now routinely plan controlled burns that run across multiple ranch boundaries. They supported a decade-long study of the relationships among cattle, vegetation, fire, and native small mammals, run on 9,000 acres of the Diamond A Ranch, and are replacing their old-fashioned fence lines with new barriers that allow wildlife to pass but still contain cattle. Some of the founding members, like Bill McDonald, heir to a century-old family ranch, have long been rotating their herds to rest pastures; the group inspired others to try such alternative ways of managing their cattle. McDonald and his neighbors have begun to keep a closer eye on their cows, moving vulnerable mothers with young calves out of the reach of mountain lions rather than trying to hunt every passing cat.

Across the western rangelands, ranchers are experimenting with new strategies. Some are bundling their herds into bigger groups of 500 or so animals, grazing one area intensely for a brief time and then moving the cattle, in imitation of once-vast swarms of highly mobile bison. In wolf country, ranchers have begun to use riders on horseback, working with herding dogs, to gather herds and protect them at night. Ecologist David Western describes this trend as the recowboying of American cattle culture. Tended and kept on the move, the impacts of the domestic cow, he says, become more like that of bison or wildebeest. Western sees important parallels between the megafauna-rich savannas of eastern Africa and the cow-dominated range of the American West.

Founder of the African Conservation Centre and former head of the Kenya Wildlife Service, Western has spent decades studying the slow ecological minuet performed by Maasai herders and their cattle as they share grasslands with elephant, giraffe, wildebeest, zebra, and other wild megafauna. In the 1970s, mainstream conservationists planned to protect the wildlife of what is now Kenya's Amboseli National Park by forcing the Maasai and their cattle out of the area. Western had a radically different vision, to protect traditional Maasai culture—cattle and all—as a way of keeping the wild megafauna safe. Having herded alongside the Maasai, Western had gained a cow's-eye view of the ecosystem. He sees the local people, and their livestock, as essential to the survival of Amboseli's wildlife.[32]

The Maasai offer a window into an alternate kind of post-Pleistocene reality, one where people and megafauna have successfully coexisted for thousands of years. The indigenous herders of east Africa tamed the aurochs as much as 9,000 years bp. Unlike most masters of domesticated beasts, they did not proceed to build and occupy permanent settlements. The Maasai roamed with their herds, shadowing the migrations of zebra and wildebeest as they sheltered in Amboseli's permanent swamps during the dry season and then spread out over the plains in times of rain. Traditional Maasai are deeply attuned to the movements of wildlife and use them to guide their herds. They know, for example, that elephants prune back the thickest swamp growth in times of drought, opening new grazing areas for zebra, wildebeest, and cattle.

Since the 1970s, however, more and more Maasai have given up the traditional life of mobile herding and now dwell in permanent huts. This trend was started by government policies that encouraged subdivision of commonly held lands. In the 1960s, conventional conservation wisdom held that the Maasai's roaming herds were overstocked, degrading the range and Amboseli's fever-tree woodlands. Settled, commercial ranching, it was thought, would be far more efficient. The Maasai rejected the idea at first—they knew they could not survive dry seasons without moving their herds to follow the availability of water and fresh grass. But the Maasai are a small minority, and their communally held lands have often been appropriated by outsiders. As East Africa's human population grows, Maasai people are subdividing their lands and settling down, for fear of otherwise losing everything.

Before the arrival of Europeans, Maasai people and their cattle traded places with elephant herds over time, creating a patchwork of diverse habitats: woodland, savanna, grassland, and marsh. "Cattle create trees, elephants create grasslands," according to a Maasai proverb. Using the techniques of modern ecology, Western has demonstrated the truth of that old saying and shown that the ancient system was sustainable because elephants and livestock each eat the plants encouraged by the other.[33] Conversely, separating the two can have disastrous effects. Since Western first came to Amboseli in the 1960s, the park's stately groves of fever-trees have vanished. The cause of the woodland decline was debated for decades until he proved elephants responsible by tracking a rebirth of fever-trees in a series of elephant exclosures in Amboseli. Grazers, like cattle and zebra, encourage the growth of fever-tree seedlings by cropping competing herbs short; elephants browse the taller trees, opening up new grasslands. A surge in ivory poaching in the 1970s forced

Fig. 32 Elephants in Amboseli National Park, Kenya, once routinely traded feeding grounds with Maasai herders and their cattle. Now elephants are confined within the park while traditional herders and cattle are banished, leading to a drop in biodiversity both inside and outside park boundaries. (Photo from Wikimedia Commons, http://en.wikipedia.org/wiki/File:Elephants_Kili_2.jpg.)

elephants inside the park, at the same time that new regulations banished cattle outside park boundaries. Today the ancient interchange of elephants, herders, and cattle has been severed in Amboseli. As a result, the national park is treeless, while the people living around its edges must hack back fever-tree thickets to plant their crops. The same pattern echoes throughout eastern Africa, and biodiversity suffers.

Though poaching is now better controlled, human settlements discourage elephants from venturing outside protected areas like Amboseli. The trend toward permanent settlements, says Western, poses the single greatest threat to East Africa's savannas and wild megafauna. He has documented a dramatic decline in wildlife populations on a section of Maasai land that was subdivided and fenced in the 1980s. On a neighboring section still occupied by traditional, mobile herders, wildlife numbers increased over the same span of time.[34] On both tracts of land, human numbers grew steadily over the thirty-three years of Western's study. The simple human act of parking in one place makes all the difference.

Permanent homes and fenced pastures block the normal movements of zebra, wildebeest, and elephant, while cattle kept year-round in one spot overgraze and deplete the forage that they once shared with wildlife. As more and more people opt for a sedentary life, the future looks grim for the Kenya–Tanzania Rift Valley area, Africa's richest remaining wildlife habitat.

Modern pressures—including a boom in the human population—are transforming Maasailand into something that begins to resemble the fragmented range of the western United States. What that could mean became clear to six Maasai who traveled to the United States to visit the Malpai Borderlands in the spring of 2004. "To our eyes," they wrote, "the Borderlands seemed bereft of wildlife. During our visit we saw three lone antelope on the range: many more of them were displayed as trophies on the walls in the houses."[35] The Maasai traditionally keep their cattle close, naming each cow, singing to her, milking her every day. At night the livestock are gathered at the center of a boma—a moveable corral in which they are surrounded by resting people, who keep watch for lions. A boma is moved with the seasons, and months after it has been uprooted, wildlife will be drawn to a rich growth of grass fertilized by cattle dung. To Maasai eyes, the Malpai way of ranching is inside out: unloved cattle, valued only for the income they generate, sleeping unprotected on the range.

The vast cultural divide between East Africa and the American Southwest did not prevent the two groups of cowboys from connecting and learning from one another. This mutual exchange, in which people from each culture visited the other's ranches, was arranged by Western and Charles Curtin, an ecologist studying cattle and wildlife in the Malpai Borderlands.[36] Communal grassbanks—a Maasai version of the shared pastures on the Diamond A Ranch—may help to keep lands open for both wildlife and cattle around Amboseli. Perhaps most important, the visit showed the Maasai in stark terms the level of wildlife loss that can result from subdividing the landscape.

It is harder to imagine carrying the lessons of Maasailand back to the United States. The Maasai have integrated themselves with the wildlife around them, maneuvering their cattle around herds of zebra and elephant with consummate skill. A lion that might contemplate a visiting tourist as a snack will flee at the sight of a Maasai. Despite the vast numbers of wildlife, face-to-face conflicts with animals are few, because the Maasai have mastered the process of living among them. It is a life completely alien to North Americans. "What you call the wild in the U.S.," says Western, "is home to the Maasai. It's the oldest cattle ranch on earth." For all the impala, giraffe, and other fantastic beasts, most of the biomass of megafauna in Amboseli—about 65 percent—consists of cattle.[37]

Western has worked with Paul Martin, and he agrees that elephants could play an important role in fighting the brush invasion in the rangelands of the southwestern United States. Closely managed herds of elephants could be moved around the region, knocking down the mesquite and juniper just as they have wiped out the fever-trees of Amboseli. In Western's view, these elephants should come from American zoos or refuges—removing wild elephants from Africa and transplanting them in the Great Plains would be both morally wrong and ecologically misguided. African wildlife is the main attraction for the continent's tourist industry, and American appropriation of that wildlife would be a kind of biopiracy. Conservation-oriented elephants in the United States should be carefully controlled—Western believes the native plants that have come to dominate the West over the last 12,000 years are ill-equipped to survive a sudden appearance of free-roaming surrogates for the long-lost megafauna.

After decades spent pondering the worldwide ecological decline wrought by the loss of megafauna, Western remains upbeat. His optimism rests on a faith in human flexibility and in the endurance of the domestic cow. The Maasai experience proves that the beast is tough, drought-hardy, and highly adaptable. Its environmental imprint depends entirely on the way it is managed. Handled in the right way, cattle can even decrease the volume of greenhouse gases released from grasslands, as demonstrated in a recent study on the Mongolian steppe.[38] We cannot raise the aurochs, but its tamed descendant may yet fill a vital ecological niche. Both our dreams and our fates remain tied to the giants around us—the humble and domesticated along with the fierce and the wild.

EPILOGUE: THE GIANTS THIS TIME

One moonless night, alone on a remote northern California ridge, I came face to face with an angry mountain lion. I had been hiking along, hooting like an owl, hoping to get some real birds to answer me. My cheap imitations annoyed the big cat—she came crashing uphill, snapping the brush that stood in her path, and erupted onto the open ridgetop, where her eyes glowed in the light from my headlamp. My hand shook as I focused the beam in her direction. The cougar circled me, hissing loudly, for what felt like an hour (actual time elapsed was more likely a minute or two). Then she subsided into the night. I sank onto a boulder, letting my panic drain away. As my heartbeat slowed toward normal, I felt a wave of euphoria—a high from my close contact with a wild creature capable of killing me. Years later, I can still hear the sibilance of the cougar's voice, like a house cat connected to a subwoofer, and see the stars that seemed to leap out of the night sky toward me when I realized she had decided to let me live.

Despite our high technology, our clouds of fossil fuel fumes and our endless miles of fence line, perhaps we are not so different from the Ice Age artists who made portraits of cave lions, mammoths, and aurochsen. A fascination with large animals is an ancient and abiding human trait, a part of our nature that may help us survive our own excesses. Many of us still yearn for the vision portrayed in Pleistocene cave paintings, a time when people were hunter-gatherers, one more free-roaming creature under a wide blue sky. No one had even dreamt of taming the towering wild ox, and a mammoth was simply a mammoth—not a symbol or a bioengineering project.

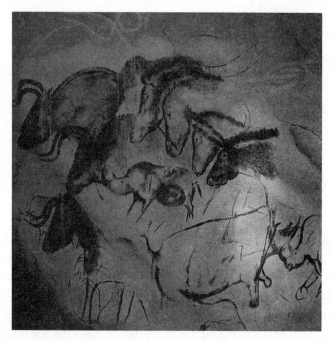

Fig. 33 Woolly rhinoceros, wild horse, and aurochs in a replica of a Pleistocene cave painting from Chauvet, France. (Photo: Wikimedia Commons http://en.wikipedia.org/wiki/File:Chauvet_cave,_paintings. JPG.)

We can never go back to those times, but understanding our long and complicated relationship with megafauna can help us go forward. Recognizing the loss of the ground sloth, the saber-toothed cat, and the short-faced kangaroo makes it possible to see today's ecosystems whole and to seek ways to keep them functional. The strategies that emerge from this understanding will often be less romantic than our Pleistocene dreams: finding the right way to move a herd of cattle through time and space, for instance, or coming to terms with managing some introduced species in cherished native habitats. We have altered the face of the planet more than our Ice Age forebears ever could have imagined. But like them, we live in a time of drastic change, and we must adapt.

A decline in wild megafauna is as much a consequence of human technologies as is global warming from man-made greenhouse gases. In fact, the two phenomena are flip sides of the same coin. Thousands of years after the first tame oxen were harnessed to pull a plow, humanity found a way to unleash a seemingly endless source of power buried in the earth. We began to mine and burn fossil fuels, releasing the ancient solar energy stored in coal and oil. At the dawn of the industrial revolution, the biomass of megafauna—no longer zebras or wolves, but people and their livestock—began to skyrocket beyond a baseline level that had held for millions of years. That trend continues today.

Anthony Barnosky, the University of California–Berkeley paleontologist, argues that we began long ago to usurp the share of global energy that once sustained mammoths and other lost megafauna. He has charted the plummeting number of native species of large wildlife over time against the steadily rising curve of human numbers, two lines that cross about 5,000 calendar years ago, during the long, slow rise of agriculture.[1] For a while, extinctions of wild megafauna stabilized, but the human population kept on growing.

Every ecosystem is powered by the sun. Solar energy captured by green plants feeds herbivores, and ultimately their predators, too. The amount of that captured energy is known to ecologists as net primary productivity (NPP), and global NPP limits the size and number of living creatures the earth can sustain. Governed by the amount of sunlight striking the planet, NPP stays relatively constant over long periods of time. So does the global net weight, or biomass, of megafauna.

During the last Ice Age, as people colonized the far reaches of the planet, their arrival coincided with a crash in the biomass of native megafauna. At first this loss was roughly balanced out by the increase in human numbers. (Then and now, we count as megafauna, defined by Barnosky and many other researchers as creatures weighing at least 44 kg [97 lbs] at adulthood. The average Stone Age person weighed 50 kg [110.2 lbs]; today the average adult weighs in at 67 kg [148 lbs].) Around the time that Clovis people spread across North America, global biomass of large animals plummeted. It took about 10,000 years after the mass extinctions before humans, together with our domesticated cattle, sheep, horses, and camels, accounted for the same amount of biological weight as the Pleistocene megafauna in their hey-day.

Then the industrial revolution broke an ancient balance, and the combined biomass of people and livestock built higher than ever. The combustion of fossil fuel has pumped abnormally high levels of carbon dioxide and other greenhouse gases into the atmosphere, creating the global warming crisis. But it has done more: it has subsidized us with an extra source of energy, allowing us to multiply out of control. Stone Age people could use only the energy contained in the prey they hunted and the plants they gathered, which imposed a natural limit on their numbers. Today, a typical U.S. citizen burns more than 100 times as much energy each year as our Pleistocene ancestors did.[2] We have accomplished this by claiming more than 40 percent of the NPP the land yields, displacing native wildlife of all kinds in the process.

The added blast of power from fossil fuels has allowed human populations to reach unprecedented peaks, an artificial boom that now teeters on the brink of collapse. The catch, of course, is that eventually oil, coal, and gas supplies will be exhausted. If we do not find alternative energy sources and reduce greenhouse gas pollution, we will suffer along with the polar bear and the caribou. If the planet experiences another crash in the biomass of megafauna like the one at the close of the last Ice Age, the human toll will be unimaginable.

The strength of domesticated giants—the pig, goat, cow, and horse—allowed us to multiply, and our numbers pushed us toward strange and powerful inventions. Neither we nor the wild creatures around us have had much time to adapt to the results. Now it is up to us to find ways to carry megafauna with us into the future, a task that will demand new ways of thinking about nature and our own place in the world.

If the people who see megafauna as indispensible shapers and protectors of life sometimes sound like dreamers, it's because they have tapped into something ancient, compelling, and completely out of context in the modern world. They've joined hands with the artists who long ago made their way deep into caves at Chauvet and Lascaux, carrying bags of pigment and lamps fueled with mammoth fat. Those Ice Age people painted pictures that detail the most effective places to spear a horse or an ibex, but that also radiate awe and delight in the living animals. They noticed the way mammoths traveled together, how aurochsen licked their newborn calves, the fury of reindeer in rut. Mixed in with all this, they left behind images of themselves. Pleistocene people understood the same thing that modern ecology reveals: the megafauna is us.

ACKNOWLEDGMENTS

Paula Levy planted the seed of this book long ago when she read to me from T.H. White's translation of a medieval bestiary. The detailed descriptions of the satyr and the manticore showed me that people have always needed big, fearsome beasts. If real ones aren't handy, we conjure some. Fly free with the griffins, sister.

The seed germinated decades later, when I worked on magazine stories about shifting visions of the Pleistocene extinctions. Thanks to Kate Douglas, my editor at *New Scientist*, and to George Black and Douglas Barasch, my editors at *OnEarth*, for sharpening those first pieces and launching me on the adventure.

My agent, Mike Hamilburg, and Peter Prescott, my first editor at Oxford, embraced this project when it was still in its earliest stages. I'm indebted to Tisse Takagi for her skillful and humane editing of the manuscript, to Warren Allmon for his insightful comments and support of my proposal, and to Trish Watson for her excellent copyediting. I also owe much to a number of talented and generous colleagues, who guided me into the world of publishing from where I started, clueless in the remote redwoods of Humboldt County. I've learned from many past and present members of the National Association of Science Writers freelance listserv, and I thank them all. Katherine Austin, Candis Condo, and T. Delene Beeland read and commented on early drafts and encouraged me when I needed it most. Tanya Kucak came up with the title when I was completely stuck. Dan Ferber, Jeff Hecht, Sandy Becker, Joel Shurkin, and many others have given excellent advice. Wendee Holtcamp is a valued comrade-in-inkstained-arms.

I am deeply grateful to the many scientists and field researchers who shared their time, insights, and adventures with me. David Bowman, Douglas Smith, the late Paul Martin, Mary Kay O'Rourke, Gary Haynes, William Ripple, David Burney, and Guy Robinson have been particularly generous and have opened my eyes in many ways. In addition to all those mentioned in the text, I thank Dennis Jenkins, Ted Goebel, Michael Collins, Samuel Wasser, David Meltzer, René Vellanoweth, Jacqui Codron, Judith Field, Steve Webb, Ian Barnes, Paul Koch, Camille Parmesan, and Noah Rosenberg.

Hugh Scanlon, husband, friend, firefighter, forester, trombonist, and wiseass, has made this book, and so much else about my life, possible. He read and offered good ideas on every draft of everything. Maya Scanlon helps me to keep all my stories simple and real. My parents, Sol and Florence Levy, raised all their children with a love of reading, writing, and wordplay.

I would have become a lifeless fossil long before I had the chance to write this book if not for the support and friendship of Catherine Finney, Rachel Hatchimonji, Dmitri Belser, Tom White, David Levy, Berk Snow, Suzanne Beers, Denise Fitzgerald, John Gullam, and that medical mensch, Bruce Kessler.

NOTES

PROLOGUE

1   Koch, Paul L., and Anthony D. Barnosky (2006). "Late Quaternary extinctions: state of the debate." *Annual Review of Ecology, Evolution, and Systematics* 37: 215–50; Martin, Paul S., and Richard G. Klein, eds. (1984). *Quaternary extinctions: a prehistoric revolution.* Tucson: University of Arizona Press, 892.

2   Miller, Gifford H., John W. Magee, Beverly J. Johnson, Marilyn L. Fogel, Nigel A. Spooner, Malcolm T. McCulloch, and Linda K. Ayliffe (1999). "Pleistocene extinction of Genyornis newtoni: human impact on Australian megafauna." *Science* 283: 205–8.

3   Donlan, Josh, Harry Greene, Joel Berger, Carl Bock, Jane Bock, David Burney, James Estes, Dave Foreman, Paul S. Martin, Gary Roemer, Felisa Smith, and Michael Soule (2005). "Re-wilding North America." *Nature* 436: 913–14.

ELEGY FOR THE MASTODON

READING THE ASHES

1   Lister, Adrian, and Paul Bahn (2007). *Mammoths: Giants of the Ice Age.* Berkeley: University of California Press.

2   Murphy, Dennis C. "Thomas Jefferson Fossil Collection, Academy of Natural Sciences, Philadelphia." http://www.ansp.org/museum/jefferson/otherPages/degeneracy-1.php

3   Robinson, G. S., Lida Pigott Burney and David A. Burney (2005). "Landscape paleoecology and megafaunal extinction in southeastern New York state." *Ecological Monographs* 75(3): 295–315.

4   Gill, Jacquelyn L., John W. Williams, Stephen T. Jackson, Katherine B. Lininger, and Guy S. Robinson (2009). "Pleistocene megafaunal collapse, novel plant communities and enhanced fire regimes in North America." *Science* 326: 1100–1103.

5   Burney, David A., Guy S. Robinson, and Lida Pigott Burney (2003). "*Sporormiella* and the late Holocene extinctions in Madagascar." *Proceedings of the National Academy of Sciences USA* 100: 10800–10805.

6   Burney, David A., and Lida Pigott Burney (1994). "Holocene charcoal stratigraphy from Laguna Tortuguero, Puerto Rico, and the timing of human arrival on the island." *Journal of Archaeological Science* 21: 273–281.

7   Steadman, David W., Paul S. Martin, Ross D. E. MacPhee, A. J. T. Jul, H. Gregory McDonald, Charles A. Woods, Manuel Iturralde-Vinent, and Gregory W. L. Hodgins (2005). "Asynchronous extinction of late Quaternary sloths on continents and islands." *Proceedings of the National Academy of Sciences USA* 102: 11763–11768.

8   White, Arthur W., Trevor H. Worthy, Stuart Hawkins, Stuart Bedford, and Matthew Spriggs (2010). "Megafaunal meiolaniid turtles survived until early human settlement in Vanuatu, Southwest Pacific." *Proceedings of the National Academy of Sciences USA. www.pnas.org/cgi/doi/10.1073/pnas.1005780107*

9   Diamond, Jared M. (2000). "Blitzkrieg against the Moas." *Science* 287: 2170–2171.

CONJURING THE GIANTS

1   Paul Martin died on September 13, 2010 at his home in Tucson. He was 82.

2   Martin, Paul S. (1961). "Rampart cave coprolite and ecology of the Shasta ground sloth." *American Journal of Science* 259: 102–127, at p. 123.

3   Ibid, at p. 125.

4   Martin, Paul S. (1966). "Africa and Pleistocene overkill." *Nature* 212: 339–344, at *p.* 341.

5   Martin, Paul S. (1967). "Prehistoric overkill." In *Pleistocene Extinctions: The Search for a Cause,,* Volume 6 of the *Proceedings of the VII Congress of the International Association for Quaternary Research*, ed. P.S. Martin and H.E. Wright, Jr., pp. 75–120. New Haven: Yale University Press.

6   Ibid, at p. 115.

7   Grayson, Donald K. (1977). "Pleistocene avifaunas and the overkill hypothesis." *Science* 195.

8   Mosimann, James E. and Paul S. Martin (1975). "Simulating overkill by Paleoindians." *American Scientist* 63: 304–313.

9   Grayson, Donald K., and David J. Meltzer. (2003). "A requiem for North American overkill." *Journal of Archaeological Science* 30: 585–593, at p. 591.

LEAVES OF GRASS

1   Janzen, Daniel H., and Paul S. Martin (1982). "Neotropical anachronisms: the fruits the gomphotheres ate." *Science* 215: 19–27.

2   Ibid.

3   Barlow, Connie C. (2000). *The Ghosts of Evolution: Nonsensical Fruit, Missing Partners, and Other Ecological Anachronisms.* New York: Basic Books.

4   Guimaraes, Paulo R., Jr., Mauro Galetti, and Pedro Jordano (2008). "Seed dispersal anachronisms: rethinking the fruits extinct megafauna ate." *PLoS One* 3(3). e1745 http://www.plosone.org/article/info%3Adoi%2F10.1371%2Fjournal.pone.0001745

5   Galetti, Mauro, Camila I. Donatti, Marco Aurelio Pizo, and Henrique C. Giacomini (2008). "Big fish are the best: seed dispersal of *Bactris glaucescens* by the pacu fish in the Pantanal, Brazil." *Biotropica* 40: 386–89.

6   Fragoso, Jose M. V., Kirsten M. Silvius, and Jose A. Correa (2003). "Long-distance seed dispersal by tapirs increases seed survival and aggregates tropical trees." *Ecology* 84: 1998–2006.

7   Wright, Joseph S., Horacio Zeballos, Ivan Dominguez, Marina M. Gallardo, Marta C. Moreno, and Roberto Ibanez (2000). "Poachers alter mammal abundance, seed dispersal, and seed predation in a neotropical forest." *Conservation Biology* 14: 227–239.

8   Maisels, F., E. Keming, M. Kemei, and C. Toh (2001). "The extirpation of large mammals and implications for montaine forest conservation: the case of the Kilum-Ijim Forest, Northwest Province, Cameroon." *Oryx* 35: 322–331.

9   Waldram, Matthew S., William J. Bond, and William D. Stock (2008). "Ecological engineering by a mega-grazer: white rhino impacts on a South African savanna." *Ecosystems* 11: 101–112

10 Western, David (2002). *In the Dust of Kilimanjaro*. Washington, DC: Island Press/Shearwater Books.

MAMMOTH TRACKS

ICE AGE DIARIES

1   Moss, Cynthia (2000). *Elephant Memories: Thirteen Years in the Life of an Elephant Family*. Chicago: University of Chicago Press.

2   Fisher, Daniel C. (2008). Paleobiology and Extinction of Proboscideans in the Great Lakes Region of North America. In American Megafaunal Extinctions at the End of the Pleistocene, ed. G. Haynes, 55–75. Dordrecht: Springer.

3   Haynes, Gary (2006). "Mammoth landscapes: good country for hunter-gatherers." *Quaternary International* 142–143: 20–29.

4   Meredith, Martin (2003). *Elephant Destiny: Biography of an Endangered Species in Africa*. New York: PublicAffairs.

5   McNeil, Paul, L. V. Hills, B. Kooyman, and Shayne M. Tolman (2005). "Mammoth tracks indicate a declining late Pleistocene population in southwestern Alberta, Canada." *Quaternary Science Reviews* 24: 1253–1259.

6   Owen-Smith, Norman (1989). "Megafaunal extinctions: the conservation message from 11,000 years bp." *Conservation Biology* 3: 405–412.

7   Pringle, Robert M. (2008). "Elephants as agents of habitat creation for small vertebrates at the patch scale." *Ecology* 89: 26–33.

8   Allmon, W. D., and P. Nester, eds. (2008). Special issue titled "Mastodon paleobiology, taphonomy, and paleoenvironment in the late Pleistocene of New York State: studies on the Hyde Park, Chemung, and North Java sites." *Palaeontographica Americana* 61.

9   Fox, David L., Daniel C. Fisher, Sergey Vartanyan, Alexei N. Tikhonov, Dick Mol, and Bernard Buigues (2007). "Paleoclimatic implications of oxygen isotopic variation in late Pleistocene and Holocene tusks of *Mammuthus primigenius* from northern Eurasia." *Quaternary International*, 169–170: 154–165.

10 Slotow, Rob, Gus van Dyk, Joyce Poole, Bruce Page, and Andre Klocke (2000). "Older bull elephants control young males." *Nature* 408: 425–26.

OF MAMMOTHS AND MEN

1   Moss, Cynthia (2001). "The demography of an African elephant population in Amboseli, Kenya." *Journal of Zoology London* 255: 145–156.

2   McComb, Karen, Cynthia Moss, Sarah M. Durant, Lucy Baker, and Soila Sayialel (2001). "Matriarchs as repositories of social knowledge in African elephants." *Science* 292: 491–494.

3   Ibid, page 494.

4   Bradshaw, Gay A., Allan N. Schore, Janine L. Brown, Joyce H. Poole, and Cynthia J. Moss (2005). "Elephant breakdown." *Nature* 433: 807, at p. 807.

5   Siebert, Charles (2006). "An elephant crackup?" *New York Times*, Oct. 8.

6   Lister, Adrian, and Paul Bahn (2007). *Mammoths: Giants of the Ice Age.* Berkeley: University of California Press.

7   Guthrie, Dale (2001). "Origin and causes of the mammoth steppe: a story of cloud cover, woolly mammal tooth pits, buckles, and inside out Beringia." *Quaternary Science Reviews* 20: 549–574.

8   Guthrie, Dale (2006). "New carbon dates link climatic change with human colonization and Pleistocene extinctions." *Nature* 441: 207–209.

9   Guthrie, Dale (2003). "Rapid body size decline in Alaskan Pleistocene horses before extinction." *Nature* 426: 169–171.

10 Hoffecker, John F., and Scott A. Elias (2007). *Human Ecology of Beringia.* New York: Columbia University Press.

11 Stuart, Anthony J., Leopold D. Sulerzhitsky, Lyobov A. Orlova, Yaroslav V. Kuzmin, and Adrian M. Lister (2002). "The latest woolly mammoths in Europe and Asia: a review of the current evidence." *Quaternary Science Reviews* 21: 1559–1569.

12 Barnes, Richard F. W. (1999). "Is there a future for elephants in West Africa?" *Mammal Review* 29: 175–199.

13 Ibid.

14 Wasser, Samuel K., Celia Mailand, Rebecca Booth, Benezeth Mutayoba, Emily Kisamo, Bill Clark, and Matthew Stephens (2007). "Using DNA to track the origin of the largest ivory seizure since the 1989 trade ban." *Proceedings of the National Academy of Sciences USA* 104: 4228–4233.

15 Gillson, Lindsey, and Keith Lindsay (2003). "Ivory and ecology—changing perspectives on elephant management and the international trade in ivory." *Environmental Science and Policy* 6: 411–419.

16 Caughley, Graeme (1976). "The elephant problem—an alternative hypothesis." *East Africa Wildlife Journal* 14: 265–283.

17 Gillson, Lindsey (2004). "Testing non-equilibrium theories in savannas: 1400 years of vegetation change in Tsavo National Park, Kenya." Ecological Complexity 1: 281–298.

18 Skarpe, C., Aarrestad P. A., Andreassen H. P., Dhillion S. S., Dimakatso T., du Toit J. T., Duncan, Halley J., Hytteborn H., Makhabu S., Mari M., Marokane W., Masunga G., Ditshoswane M., Moe S. R., Mojaphoko R., Mosugelo D., Motsumi S., Neo-Mahupeleng G., Ramotadima M., Rutina L., Sechele L., Sejoe T. B., Stokke S., Swenson J. E., Taolo C., Vandewalle M., Wegge P. (2004)."Return of the giants: ecological effects of an increasing elephant population." Ambio 33: 276–82.

19 Cerling, Thure E., George Wittemyer, Henrik B. Rasmussen, Fritz Vollrath, Claire E. Cerling, Todd J. Robinson, and Iain Douglas-Hamilton (2006). "Stable isotopes in elephant hair document migration patterns and diet changes." *Proceedings of the National Academy of Sciences USA* 103: 371–373.

20 van Aarde, Rudi J., and Tim P. Jackson (2007). "Megaparks for metapopulations: addressing the causes of locally high elephant numbers in southern Africa." *Biological Conservation* 134: 289–297.

FIRST ENCOUNTERS

1 Haynes, C. Vance, Jr., and Bruce B. Huckell, eds. (2007). *Murray Springs: A Clovis Site with Multiple Activity Areas in the San Pedro Valley, Arizona*. Anthropological Papers of the University of Arizona 71. Tucson: University of Arizona Press.

2 Haynes, C. Vance (1964). "Fluted projectile points: their age and dispersion." *Science* 145: 1408–1413.

3 At Paisley Caves in central Oregon, archaeologist Dennis Jenkins discovered 14,000-year-old human feces, along with distinctly non-Clovis stone tools and bones of Pleistocene camel and horse. The coprolites were found to contain human DNA with sequences unique to Native Americans. This unexpected scatological proof of a pre-Clovis human presence in the Americas set off a firestorm of scientific debate. Still, for many researchers, it clinches the argument that Clovis was not first. See M. Thomas P. Gilbert, Dennis L. Jenkins, Anders Gotherstrom, Nuria Naveran, Juan J. Sanchez, Michael Hofreiter, et al. (2008). "DNA from pre-Clovis human coprolites in Oregon, North America." *Science* 320: 786–89; and Balter, Michael (2008). "DNA from fossil feces breaks Clovis barrier." *Science* 320: 37.

4 Dillehay, Thomas D. (2000). *The Settlement of the Americas: A New Prehistory*. New York: Basic Books.

5 Dillehay, Tom D., C. Ramirez, M. Pino, M. B. Collins, J. Rossen, J. D. Pino-Navarro (2008). "Monte Verde: seaweed, food, medicine and the peopling of South America." *Science* 320: 784–86.

6 Martin, Paul S. (2005). *Twilight of the Mammoths: Ice Age Extinctions and the Rewilding of America*, at p. 138. Berkeley: University of California Press.

7 Pitulko, V. V., P. A. Nikolsky, E. Yu. Girya, A. E. Basilyan, V. E. Tumskoy, S. A. Koulakov, S. N. Astakhov, E. Yu. Pavlova, and M. A. Anisimov (2004). "The Yana RHS site: humans in the Arctic before the last glacial maximum." *Science* 303: 52–56.

8 Hoffecker and Elias, *Human Ecology of Beringia*.

9 Goebel, Ted (2009). Personal communication, Texas A&M University.

10 Dixon, James E. (1999). *Bones, Boats and Bison: Archeology and the First Colonization of Western North America*. Albuquerque: University of New Mexico Press.

11 Hoffecker and Elias, *Human Ecology of Beringia*.

12 Dixon, James E. (2001). "Human colonization of the Americas: timing, technology and process." *Quaternary Science Reviews* 20: 277–299.

13 Dandridge, Debra E. (2006). "First lady of the new world: Arlington Springs woman." *Mammoth Trumpet* 21: 4–6.

14 Schroeder, K. B., T. G. Schurr, J. C. Long, N. A. Rosenberg, M. H. Crawford, L. A. Tarskaia, L. P. Osipova, S. I. Zhadanov, and D. G. Smith (2007). "A private allele ubiquitous in the Americas." *Biology Letters* 3: 218–223.

15 Vellanoweth, Rene (2009). Personal communication, California State University–Los Angeles.

16 Haynes, C. Vance (2008). "Younger Dryas black mats and the Rancholabrean termination in North America." Proceedings of the National Academy of Sciences USA 105: 6520–6525.

17 Recently, a group of researchers claimed that the demise of mammoths, mastodons, and Clovis people were all triggered by the impact of a comet striking the earth's atmosphere. They believe this resulted in a firestorm that torched forests and grasslands over much of the continent [Firestone, R. B., A. West, J. P. Kennett, et al. (2007). "Evidence for an extraterrestrial impact 12,900 years ago that contributed to the megafaunal extinctions and the Younger Dryas cooling." Proceedings of the National Academy of Sciences USA 104: 16016–16021]. The hypothesis spawned some sensational press coverage, but the science supporting it has been debunked by Haynes and others [Surovell, Todd A., Vance T. Holliday, Joseph A.M. Gingerich, Caroline Ketron, C. Vance Haynes, Jr., Ilene Hilman, Daniel P. Wagner, Eileen Johnson, and Philippe Claeys (2009). "An independent evaluation of the Younger Dryas extraterrestrial impact hypothesis." Proceedings of the National Academy of Sciences USA 106: 18155–18158; Kerr, Richard (2008). "Experts find no evidence for a mammoth-killing impact." Science 319: 1331–1332]; Scott, A. C., N. Pinter, M. E. Collinson, M. Hardiman, R. S. Anderson, A. P. R. Brain, S. Y. Smith, F. Marone, and M. Stampanoni (2010) Fungus, not comet or catastrophe, accounts for carbonacious spherules in Younger Dryas "impact layer". Geophysical Research Letters 37 doi:10.1029/2010GL043345

GIANTS DOWN UNDER

PREHISTORIC PITFALLS

1 Wells, R. T., K. Moriarty, and D. L. G. Williams (1984). "The fossil vertebrate deposits of Victoria Fossil Cave, Naracoorte: an introduction to the geology and fauna." *Australian Zoologist* 21(4): 305–33.

2 Owen, F. R. S. (1870). "On the fossil mammals of Australia, Part III. *Diprotodon australis*, Owen." Philosophical Transactions of the Royal Society of London, 160: 519–78.

3 Price, G. J. (2008). "Taxonomy and palaeobiology of the largest-ever marsupial, *Diprotodon* Owen, 1838 (Diprotodontidae, Marsupialia)." *Zoological Journal of the Linnean Society* 153: 369–97.

4 Wells, R. T., D. R. Horton, and P. Rogers (1982). "Thylacoleo carnifex Owen (Thylacoleonidae): marsupial carnivore?" p. 573–85 in *Carnivorous Marsupials*, M. Archer, ed., Royal Zoological society of New South Wales, Sydney, Australia.

5   Wells, R. T., and B. Nichol (1977). "On the manus and pes of *Thylacoleo car-nifex* Owen (Marsupialia)." *Transactions of the Royal Society of South Australia* 101(6): 139–46.

6   Prideaux, Gavin J., Richard G. Roberts, Dirk Megirian, Kira E. Westaway, John C. Hellstrom, and Jon M. Olley (2007). "Mammalian responses to Pleistocene climate change in southeastern Australia." *Geology* 35(1): 33–36.

7   In recent years there have been several claims of sites containing older human artifacts, but none have stood up to close scrutiny. See Turney, Chris S. M. Michael I. Bird, L. Keith Fifield, Richard G. Roberts, Mike Smith, et al. (2001). "Early human occupation at Devil's Lair, southwestern Australia 50,000 years ago." *Quaternary Research* 55: 3–13.

8   Roberts, Richard.G., Timothy F. Flannery, Linda K. Ayliffe, et al. (2001). "New ages for the last Australian megafauna: continent-wide extinction about 46,000 years ago." *Science* 292: 1888–92.

9   Prideaux, Gavin J., John A. Long, Linda K. Ayliffe, John C. Hellstrom, Brad Pillans, Walter E. Boles, et al. (2007). "An arid-adapted middle Pleistocene verte-brate fauna from south-central Australia." *Nature* 445: 422–25.

10   Trueman, Clive N. G., Judith H. Field, Joe Dortch, Bethan Charles, and Stephen Wroe (2005). "Prolonged coexistence of humans and megafauna in Pleistocene Australia." *Proceedings of the National Academy of Sciences USA* 102(23): 8381–85; Grün, Rainer, Stephen Egginsa, Maxime Auberta, Nigel Spoonera, Alistair W.G. Pike, and Wolfgang Müllera (2010). "ESR and U-series analyses of faunal mate-rial from Cuddie Springs, NSW, Australia: implications for the timing of the extinction of the Australian megafauna." *Quaternary Science Reviews* 29(5–6): 596–610.

11   Brook, B. W., and C. N. Johnson (2006). "Selective hunting of juveniles as a cause of the imperceptible overkill of the Australian Pleistocene megafauna." *Alcheringa: An Australasian Journal of Palaeontology* 30, Special Issue 1: 39–48.

12   Johnson, Chris (2006). *Australia's Mammal Extinctions: A 50,000 Year History.* Cambridge: Cambridge University Press.

VISIONS OF FIRE

1   Pyne, S. (1998). *Burning Bush: A Fire History of Australia.* Seattle & London: University of Washington Press.

2   Bowman, D. M. J. S., A. Walsh, and L. D. Prior (2004). "Landscape analysis of Aboriginal fire management in Central Arnhemland, North Australia." *Journal of Biogeography* 31: 207–23.

3   Miller, Gifford H., John W. Magee, Beverly J. Johnson, Marilyn L. Fogel, Nigel A. Spooner, Malcolm T. McCulloch, and Linda K. Ayliffe (1999). "Pleistocene extinction of *Genyornis newtoni*: human impact on Australian megafauna." *Science* 283: 205–8.

4   Brook, Barry.W., David M. J. S. Bowman, David A. Burney, Timothy F. Flannery, Michael K. Gagan, Richard Gillespie, Christopher N. Johnson, Peter Kershaw, John W. Magee, Paul S. Martin, Gifford H. Miller, Benny Peiser, and Richard G. Roberts (2007). "Would the Australian megafauna have become extinct if humans had never colonised the continent? Comments on 'A review of the evidence for a

human role in the extinction of Australian megafauna and an alternative explanation' by S. Wroe and J. Field." *Quaternary Science Reviews* 26: 560–64.

5 Miller, Gifford. H., Marilyn L. Fogel, John W. Magee, Michael K. Gagan, Simon J. Clarke, and Beverly J. Johnson (2005). "Ecosystem collapse in Pleistocene Australia and a human role in megafauna extinction." *Science* 309: 287–90.

6 Bowman, D. M. J. S., G. S. Boggs, L. D. Prior, and E. S. Krull (2007). "Dynamics of *Acacia aneura-Triodia* boundaries using carbon and nitrogen signatures in soil organic matter in central Australia." *Holocene* 17(3): 311–18.

7 Brook, B. W. and D. M. J. S. Bowman (2004). "The uncertain blitzkrieg of Pleistocene megafauna." *Journal of Biogeography* 31: 517–23.

8 Johnson, *Australia's Mammal Extinctions*.

9 Bowman, D. M. J. S., B. P. Murphy, and C.R. McMahon (2010). "Using carbon isotope analysis of the diet of two introduced Australian megaherbivores to understand Pleistocene megafaunal extinctions." *Journal of Biogeography* 37: 499–505.

10 Burgess, Christopher P., Faye. H. Johnston, Helen L. Berry, Joseph McDonnell, Dean Yibarbuk, Charlie Gunabarra, Albert Mileran, and Ross S. Bailie (2009). "Healthy country, healthy people: the relationship between Indigenous health status and 'caring for country.'" *Medical Journal of Australia* 190(10): 567–72.

WILD DREAMS

THE WOLF RETURNS

1 Ripple, William J., Eric J. Larsen, Roy A. Renkin, and Douglas W. Smith (2001). "Trophic cascades among wolves, elk and aspen on Yellowstone National Park's northern range." *Biological Conservation* 102: 227–34.

2 Ibid.

3 Ripple, William J., and Robert L. Beschta (2006). "Linking wolves to willows via risk-sensitive foraging by ungulates in the northern Yellowstone ecosystem." *Forest Ecology and Management* 230: 96–106.

4 Ripple, William J. and Robert L. Beschta (2007). "Restoring Yellowstone's aspen with wolves" *Biological Conservation* 138: 514–519. Not all scientists agree that a wolf-driven change in elk behavior is restoring the park's aspen stands, however. See Kauffman, Matthew J., Jedediah F. Brodie, and Erik S. Jules (2010). Are wolves saving Yellowstone's aspen? A landscape-level test of a behaviorally mediated trophic cascade. *Ecology* 91:2742–55.

5 Smith, D. W. and D. B. Tyers (2008). "Beavers of Yellowstone." *Yellowstone Science* 16(3): 4–15.

6 Hebblewhite, Mark, Clifford A. White, Clifford G. Nietvelt, et al. (2005). "Human activity mediates a trophic cascade caused by wolves." *Ecology* 86(8): 2135–44.

7 Berger, Joel., Peter B. Stacey, Lori Bellis, and Matthew P. Johnson (2001). "A mammalian predator-prey imbalance: grizzly bear and wolf extinction affect avian neotropical migrants." *Ecological Applications* 11(4): 947–60.

8   Hebblewhite, M., Clifford A. White, Clifford G. Nietvelt, et al. (2005). "Human activity mediates a trophic cascade caused by wolves." *Ecology* 86(8): 2135–44.

9   Estes, J. A., and David O. Duggins (1995). "Sea otters and kelp forests in Alaska: generality and variation in a community ecological paradigm." *Ecological Monographs* 65(1): 75–100.

10 Terborgh, J., Lawrence Lopez, Percy Nunez, Madhu Rao, et al. (2001). "Ecological meltdown in predator-free forest fragments." *Science* 294: 1923–26.

11 Levy, S. (2006). "A plague of deer." *BioScience* 56(9): 718–21.

12 Leopold, Aldo (1949). *A Sand County Almanac, and Sketches Here and There*, at p. 130. New York: Oxford University Press.

13 Ibid, at p. 196.

14 Ripple, William J. and Robert L. Beschta (2005). "Linking Wolves and Plants: Aldo Leopold on Trophic Cascades." *BioScience* 55(7): 613–21.

15 Binkley, Dan, Margaret Moore, William Romme, and Peter Brown (2006). "Was Aldo Leopold Right about the Kaibab Deer Herd?" *Ecosystems* 9:227–241; Rasmussen, D. Irvin (1941). "Biotic communities of Kaibab Plateau, Arizona." *Ecological Mongraphs* 11(3): 230–275.

16 Beschta, Robert L., and William J. Ripple (2008). "Wolves, trophic cascades, and rivers in Olympic National Park, USA." *Ecohydrology* 1: 118–30.

17 Ripple, William J., and Robert L. Beschta (2006). "Linking a cougar decline, trophic cascade, and catastrophic regime shift in Zion National Park." *Biological Conservation* 133: 397–408.

18 Kauffman, Matthew J., Nathan Varley, Douglas W. Smith, et al (2007). "Landscape heterogeneity shapes predation in a newly restored predator-prey system." *Ecology Letters* 10: 690–700.

19 Hebblewhite, Mark, and Evelyn H. Merrill (2009). "Trade-offs between predation risk and forage differ between migrant strategies in a migratory ungulate." *Ecology* 90(12): 3445–54.

20 Wright, Gregory J., Rolf O. Peterson, Douglas W. Smith, and Thomas O. Lemke (2006). "Selection of northern Yellowstone elk by gray wolves and hunters." *Journal of Wildlife Management* 70(4): 1070–78.

21 Ripple, William J. and Blaire Van Valkenburgh (2010) "Linking top-down forces to the Pleistocene megafaunal extinctions." *BioScience* 60(7): 516–26.

22 Berger, Kim M., Eric M. Gese, and Joel Berger (2008). "Indirect effects and traditional trophic cascades: a test involving wolves, coyotes, and pronghorn." *Ecology* 89(3): 818–28.

23 Glick, D. (2007). "End of the road?" *Smithsonian*, January. http://www.smithsonianmag.com/science-nature/pronghorn.html

24 Crooks, Kevin R., and Michael E. Soulé (1999). "Mesopredator release and avifaunal extinctions in a fragmented system." *Nature* 400: 563–66.

25 Anonymous (2007). "Initiative inspired by wandering wolf." *Connections: Publication of the Yellowstone to Yukon Conservation Initiative* (13): 7.

26 Donlan, Josh., Harry Greene, Joel Berger, Carl Bock, Jane Bock, David Burney, James Estes, Dave Foreman, Paul S. Martin, Gary Roemer, Felisa Smith, and Michael Soulé (2005). "Re-wilding North America." *Nature* 436: 913–14.

TALKING ABOUT A REVOLUTION

1   Deevey, E. S. (1967). "Introduction." P. 63-72, at p. 72, In: Martin, P. S. and H. E. Wright, Jr., editors, *Pleistocene Extinctions: The Search for a Cause, Volume 6 of the Proceedings of the VII Congress of the International Association for Quaternary Research.* New Haven, Yale University Press.

2   Martin, P. S. (1969). "Wanted: a suitable herbivore." *Natural History,* February: 35-39.

3   Martin, Paul S., and David A. Burney (1999). "Bring back the elephants!" *Wild Earth,* Spring: 57-64.

4   Levy, S. (2003). "Getting the drop on Hawaiian invasives." *BioScience* 53(8): 694-99.

5   Donlan, Josh. C., Karl Campbell, Wilson Cabrera, Christian Lavoie, Victor Carrion, and Felipe Cruz (2007). "Recovery of the Galapagos rail following the removal of invasive mammals." *Biological Conservation* 138: 520-24.

6   Soulé, M. E. (1990). "The onslaught of alien species, and other challenges in the coming decades." *Conservation Biology* 4(3): 233-39.

7   Morrell, V. (2009). "Research wolves of Yellowstone killed in hunt." *Science* 326: 506-7.

8   Holmern, T., Julius Nyahongo, and Eivin Roskaft (2007). "Livestock loss caused by predators outside the Serengeti National Park, Tanzania." *Biological Conservation* 135: 518-26.

9   Packer, C., Dennis Ikanda, Bernard Kissui, and Hadas Kushnir (2005). "Lion attacks on humans in Tanzania." *Nature* 436: 927-28.

10   Packer, C. (2009). "Rational fear." *Natural History Magazine,* May.

11   Marker, L. L., M. G. L. Mills and D. W. MacDonald (2003). "Factors influencing perceptions of conflict and tolerance toward cheetahs on Namibian farmlands." Conservation Biology 17(5): 1290-1298.

12   Barnett, Ross, Ian Barnes, Matthew J. Phillips, Larry D. Martin, C. Richard Harington, Jennifer A. Leonard and Alan Cooper (2005). "Evolution of the extinct sabretooths and the American cheetah-like cat." Current Biology 15(15): R589

13   Cooley, H. S., Hugh S. Robinson, Robert B. Wielgus, and Catherine S. Lambert (2008). "Cougar prey selection in a white-tailed deer and mule deer community." *Journal of Wildlife Management* 72(1): 99-106.

14   Wielgus, R. B., and F. L. Bunnell (2000). "Possible negative effects of adult male mortality on female grizzly bear reproduction." *Biological Conservation* 93: 145-54.

15   Daniel, R. MacNulty, Douglas W. Smith, John A. Vucetich, L. David Mech, Daniel R. Stahler, and Craig Packer (2009). "Predatory senescence in ageing wolves." *Ecology Letters* 12: 1-10; Rutledge, Linda Y., Brent R. Patterson, Kenneth J. Mills, Karen M. Loveless, Dennis L. Murray, and Bradley N. White (2010). "Protection from harvesting restores the natural social structure of eastern wolf packs." *Biological Conservation* 143: 332-39; Levy, S. (2010). "Family values: why wolves belong together." *New Scientist* 2764: 40-43.

16   Morrell, V. (2009). "Research wolves of Yellowstone killed in hunt." *Science* 326: 506-7.

WILD REALITIES

RANDOM ACTS OF REWILDING

1   Thomas, H. S. (1979). *The Wild Horse Controversy*. South Brunswick, NJ: Barnes.

2   Wagner, F. H. (1983). "Status of wild horse and burro management on public rangelands." In *Transactions of the 48th North American Wildlife and Natural Resources Conference*, pp. 116–33. Washington, DC: Wildlife Management. Institute.

3   Beever, E. A. (2003). "Management implications of the ecology of free-roaming horses in semi-arid ecosystems of the western United States." *Wildlife Society Bulletin* 31(3): 887–95.

4   Turner, J. W., Jr., and M. L. Morrison (2001). "Influence of predation by mountain lions on numbers and survivorship of a feral horse population." *Southwestern Naturalist* 46(2): 183–90.

5   Stacey, May Humphreys and Edward Fitzgerald Beale (1929). *Uncle Sam's Camels; The Journal of May Humphreys Stacey Supplemented by the Report of Edward Fitzgerald Beale (1857–1858)*. Cambridge, MA: Harvard University Press. Reprint, ed. L.B. Lesley. Glorieta, NM: Rio Grande Press, 1970.

6   Ibid, at p.147.

7   Huc, Regis-Evariste and Joseph Gabet (1928). *Travels in Tartary, Thibet, and China, 1844–46*. New York: Harper and Bros. As cited in Fowler, Harlan D. (1980). *Three Caravans to Yuma: the untold story of Bactrian camels in western America*, at p. 16. Glendale, California: A.H. Clark Co.

8   Barnard, W. C. (n.d. [1940s]). "Hi Jolly and the U.S. Camel Corps: Prospector Convinced Arizona Still Has Camels." *Associated Press*.

9   Edwards, G. P., B. Zeng, W. K. Saalfeld, P. Vaarzon-Morel, and M. McGregor, eds. (2008). *Managing the Impacts of Feral Camels in Australia: A New Way of Doing Business*." Report 47. Alice Springs, Australia: Desert Knowledge CRC.

10 Dorges, Birgit, and Jürgen Heucke (2003). "Demonstration of ecologically sustainable management of camels on aboriginal and pastoral land." Final report on project 200046. Canberra, Australia: Natural Heritage Trust.

11 Collis, B. (1999). "Wild things." *Ecos* 98: 6–9.

12 Tedmanson, Sophie (2009). "Feral camels terrorise Australian Outback community." *Times Online*, November 26, 2009, ww.timesonline.co.uk/tol/news/environment/article6930925.ece; National Indigenous Radio Service (2009) "Docker River camel cull nearing completion," *National Indigenous Radio News Review*, December 14, 2009. http://www.nirs.org.au/index.php?option=com_content&view=article&id=2066:docker-river-camel-cull-nearing-completion&catid=2:news-broadcasts&Itemid=65

13 Edwards et al., *Managing the Impacts of Feral Camels in Australia*.

14 Haber, G. C. (1996). "Biological, conservation, and ethical implications of exploiting and controlling wolves." *Conservation Biology* 10(4): 1068–81.

15 Thomson, P. C., K. Rose, and N. E. Kok (1992). "The behavioural ecology of dingoes in northwestern Australia. V. Population dynamics and variation in the social system." *Wildlife Research* 19: 565–84.

16   Wallach, Arian. D., Brad R. Murray, and Adam J. O'Neill (2009). "Can threat-ened species survive where the top predator is absent?" *Biological Conservation* 142: 43–52.
17   Corbett, L. K. (1995). *The Dingo in Australia and Asia.* Ithaca, NY: Comstock/ Cornell.
18   Johnson, Chris (2006). *Australia's Mammal Extinctions: A 50,000 Year History.* Cambridge: Cambridge University Press.
19   Wroe, S., C. McHenry, and J. Thomason (2005). "Bite club: comparative bite force in big biting mammals and the prediction of predatory behaviour in fossil taxa." *Proceedings of the Royal Society of London, Series B* 272: 619–25.
20   Johnson, Christopher N., Joanne L. Isaac, and Diana O. Fisher (2007). "Rarity of a top predator triggers continent-wide collapse of mammal prey: dingoes and marsupials in Australia." *Proceedings of the Royal Society of London, Series B* 274: 341–46.
21   Letnic, M., M. S. Crowther, and F. Koch (2009). "Does a top predator provide an endangered rodent with refuge from an invasive mesopredator?" *Animal Conservation* 12(4): 302–12.

WILD BY DESIGN

1   Bunce, Michael, Marta Szulkin, Heather R. L. Lerner, Ian Barnes, Beth Shapiro, Alan Cooper, and Richard N. Holdaway (2005). "Ancient DNA provides new insights into the evolutionary history of New Zealand's extinct giant eagle." *PLoS Biol* 3(1): e9.
2   Diamond, J. M. (2000). "Blitzkrieg against the moas." *Science* 287: 2170–71.
3   Anderson, Atholl, and Ian Smith (1996). "The transient village in southern New Zealand." *World Archaeology* 27(3): 359–71.
4   Bond, William J., William G. Lee, and Joseph M. Craine (2004). "Plant structural defences against browsing birds: a legacy of New Zealand's extinct moas." *Oikos* 104: 500–8.
5   Wood, Jamie R., Nicolas J. Rawlence, Geoffery M. Rogers, Jeremy J. Austin, Trevor H. Worthy, and Alan Cooper (2008). "Coprolite deposits reveal the diet and ecology of the extinct New Zealand megaherbivore moa." *Quaternary Science Reviews* 27: 2593–602.
6   Ibid.
7   Lee, William G., Jamie R. Wood, and Geoffrey M. Rogers (2010). "Legacy of avian-dominated plant-herbivore systems in New Zealand." *New Zealand Journal of Ecology* 34(1): 28–47.

THE PACE OF THE TORTOISE

1   Hansen, Dennis M., Christopher N. Kaiser, and Christine B. Muller (2008). "Seed dispersal and establishment of endangered plants on oceanic islands: the Janzen-Connell model, and the use of ecological analogues." *PLoS One* 3(5): e2111.

2   Hansen, Dennis M., and Mauro Galetti (2009). "The forgotten megafauna." *Science* 324: 42–43.

3   Truett, Joe, and Mike Phillips (2009). "Beyond historic baselines: restoring Bolson tortoises to Pleistocene range." *Ecological Restoration* 27(2): 144–51.

4   Hansen, Dennis M., C. Josh Donlan, Christine J. Griffiths and Karl J. Campbell (2010). "Ecological history and latent conservation potential: large and giant tortoises as a model for taxon substitutions." *Ecography* 33: 272–284

5   Emslie, S. D. (1987). "Age and diet of fossil california condors in Grand Canyon, Arizona." *Science* 237: 768–70.

6   Chamberlain, C. P., J. R. Waldbauer, K. Fox-Dobbs, S. D. Newsome, P. L. Koch, et al. (2005). "Pleistocene to recent dietary shifts in California condors." *Proceedings of the National Academy of Sciences USA* 102(46): 16707–11.

7   Cohn, J. P. (1993). "The flight of the California condor." *BioScience* 43(4): 206–9.

8   Grantham, J. (2010). U.S. Fish and Wildlife Service Personal communication.

THE BIG HEAT

MELTING ICE

1   Wohlforth, Charles (2004). *The Whale and the Supercomputer.* New York: North Point Press.

2   Angliss, R. P., and R. B. Outlaw (2007). "Bowhead whale (*Balaena mysticetus*): western Arctic stock." In: *Alaska Marine Mammal Stock Assessments, 2006.* NOAA Technical Memorandum NMFS-AFSC-168. Seattle, WA: U.S. Department of Commerce, National Oceanic and Atmospheric Administration, National Marine Fisheries Service, Alaska Fisheries Science Center.

3   Ibid.

4   George, J. C., J. Zeh, R. Suydam, and C. Clark (2004). "Abundance and population trend (1978–2001) of western Arctic bowhead whales surveyed near Barrow, Alaska." *Marine Mammal Science* 20: 755–773.

5   Moore, Sue, and Kristin L. Laidre (2006). "Trends in sea ice cover within habitats used by bowhead whales in the western Arctic." *Ecological Applications* 16: 932–44.

6   Amstrup, Steven C., Ian Stirling, Tom S. Smith, Craig Perham, and Gregory W. Thiemann (2006). "Recent observations of intraspecific predation and cannibalism among polar bears in the southern Beaufort Sea." *Polar Biology* 29(11): 997–1002.doi: 10.1007/s00300-006-0142-5.

7   Ibid.

8   Regehr, Eric V., Nicholas J. Lunn, Steven C. Amstrup and Ian Stirling (2007). "Effects of earlier sea ice breakup on survival and population size of polar bears in western Hudson Bay." *Journal of Wildlife Management* 71(8): 2673–83.

9   Amstrup, Steven C., Hal Caswell, Eric DeWeaver, Ian Stirling, David C. Douglas, Bruce G. Marcot, and Christine M. Hunter (2009). "Rebuttal of 'Polar bear population forecasts: a public-policy forecasting audit." *Interfaces* 39(4): 353–369; Lindqvist, Charlotte, Stephan C. Schuster, Yazhou Sun, Sandra L. Talbot, Ji Qi, Aakrosh Ratan, et al. (2010). "Complete mitochondrial genome of a Pleistocene

jawbone unveils the origin of polar bear." *Proceedings of the National Academy of Sciences* 107(11): 5053-57.

10   Grayson, Donald K., and Francoise Delpech (2003). "Ungulates and the middle to upper paleolithic transition at Grotte Xvi (Dordogne, France)." *Journal of Archaeological Science* 30: 1633-48; Grayson, Donald K., and Francoise Delpech (2005). "Pleistocene reindeer and global warming." *Conservation Biology* 19(2): 557-62.

11   Vors, Liv S., and Mark Stephen Boyce (2009). "Global Declines of Caribou and Reindeer." *Global Change Biology:* 15(11): 2626-33 15: 2626-33.

12   Post, Eric and Mads C. Forchhammer (2008). "Climate change reduces reproductive success of an Arctic herbivore through trophic mismatch." *Philosophical Transactions of the Royal Society, Series B, Biological Sciences* 363: 2369-75.

13   Weladji, Robert B., Oystein Holand, and Trygve Almoy (2003). "Use of climatic data to assess the effect of insect harassment on the autumn weight of reindeer calves." *Journal of Zoology* (London) 260: 79-85; Colman, Jonathan E., Christian Pedersen, Dag O. Hjermann, Oystein Holand, Stein R. Moe, and Eigil Reimers (2003). "Do wild reindeer exhibit grazing compensation during insect harassment?" *Journal of Wildlife Management* 67(1): 11-19.

14   Guthrie, Dale (2005). *The Nature of Paleolithic Art.* Chicago: University of Chicago Press.

15   Kuzyk, Gerald W., Donald E. Russell, Richars S. Farnell, Ruth M. Gotthardt, P. Gregory Hare, and Erik Blake (1999). "In pursuit of prehistoric caribou on Thandlat, southern Yukon." *Arctic* 52(2): 214-219.

MOVING WITH THE TIMES

1   Solomon, S., D. Qin, M. Manning, Z. Chen, M. Marquis, K. B. Averyt, M. Tignor, H. L. Miller, and Z. Chen, eds. (2007). *Climate Change 2007: The Physical Science Basis. Contribution of Working Group I to the Fourth Assessment Report of the Intergovernmental Panel on Climate Change.* Cambridge: Cambridge University Press.

2   Barnosky, Anthony D. (2009). *Heatstroke: Nature in an Age of Global Warming.* Washington, DC: Island Press/Shearwater Books.

3   Beever, Erik A., Peter F. Brussard, and Joel Berger (2003). "Patterns of apparent extirpation among isolated populations of pikas in the Great Basin." *Journal of Mammalogy* 84(1): 37-54.

4   Moritz, Craig, James L. Patton, Chris J. Conroy, et al. (2008). "Impact of a century of climate change on small-mammal communities in Yosemite National Park, USA." *Science* 322: 261-264.

5   Barnosky, Anthony D., Christopher J. Bell, Steven D. Emslie, et al. (2004). "Exceptional record of mid-Pleistocene vertebrates helps differentiate climatic from anthropogenic ecosystem perturbations." *Proceedings of the National Academy of Sciences USA* 101(25): 9297-9302.

6   Ogutu, Joseph O., and Norman Owen-Smith (2003). "ENSO, rainfall and temperature influences on extreme population declines among African savanna ungulates." *Ecology Letters* 6: 412-419.

7   Thuiller, Wilfried, Olivier Broennimann, Greg Hughes, J. Robert M. Alkemade, Guy F. Midgley, and Fabio Corsi (2006). "Vulnerability of African mammals to anthropogenic climate change under conservative land transformation assumptions." *Global Change Biology* 12: 424–440.

8   Parmesan, Camille (2006). "Ecological and evolutionary responses to recent climate change." *Annual Review of Ecology, Evolution, and Systematics* 37: 637–669.

9   McLaughlin, John F., Jessica J. Hellman, Carol L. Boggs, and Paul R. Ehrlich (2002). "Climate change hastens population extinctions." *Proceedings of the National Academy of Sciences USA* 99(9): 6070–74.

10 McLaughlin, John F., Jessica J. Hellman, Carol L. Boggs, and Paul R. Ehrlich (2002). "The route to extinction: population dynamics of a threatened butterfly." *Oecologia* 132: 538–48.

11 Mueller, Jillian M., and Jessica J. Hellmann (2008). "An assessment of invasion risk from assisted migration." *Conservation Biology* 22(3): 562–567.

12 Epps, Clinton W., Dale R. McCullough, John D. Wehausen, Vernon C. Bleich, and Jennifer L. Rechel (2004). "Effects of climate change on population persistence of desert-dwelling mountain sheep in California." *Conservation Biology* 18(1): 102–113.

13 Wilmers, Christopher C., and Wayne M. Getz (2005). "Gray wolves as climate change buffers in Yellowstone." *PLoS Biology* 3(4): e92.

14 Wilmers, Chris (2009). University of California at Santa Cruz. Personal communication.

15 Oechel, Walter C., George L. Vourlitis, Steven J. Hastings, Rommel C. Zulueta, Larry Hinzman, and Douglas Kane (2000). "Acclimation of ecosystem CO_2 exchange in the Alaskan Arctic in response to decadal climate warming." *Nature* 406: 978–81.

16 Sturm, Matthew, Josh Schimel, Gary Michaelson, Jeffrey M. Welker, Steven F. Oberbauer, Glen E. Liston, Jace Gahnestock, and Vladimir E. Romanovsky (2005). "Winter biological processes could help convert Arctic tundra to shrubland." *BioScience* 55(1): 17–26.

17 Zimov, S. A. (2005). "Pleistocene park: return of the mammoth's ecosystem." *Science* 308: 796–98.

18 Zimov, S. A., V. I. Chuprynin, A. P. Oreshko, F. S. Chapin, J. F. Reynolds, and M. C. Chapin (1995). "Steppe-tundra transition: a herbivore-driven biome shift at the end of the Pleistocene." *American Naturalist* 146: 765–94.

19 Vitebsky, Piers (2005). *The Reindeer People: Living with Animals and Spirits in Siberia*. London: HarperCollins.

20 Post, Eric, and Christian Pedersen (2008). "Opposing plant community responses to warming with and without herbivores." *Proceedings of the National Academy of Sciences USA* 105(34): 12353–358.

DEAD BEASTS WALKING

THE MAMMOTH DECIPHERED

1   Miller, Webb, Daniela I. Drautz, Aakrosh Ratan, Barbara Pusey, Ji Qi, Arthur M. Lesk, Lynn P. Tomsho, Mochael D. Packard, Fangqing Zhao, Andrei Sher,

Alexei Tikhonov, Brian Raney, Nick Patterson, Kerstin Lindblad-Toh, Eric S. Lander, James R. Knight, Gerard P. Irzyk, Karin M. Fredrikson, Tomothy T. Harkins, Sharon Sheridan, Tom Pringle, and Stephan C. Schuster (2008). "Sequencing the nuclear genome of the extinct woolly mammoth." *Nature* 456: 387–90.

2   Gilbert, Thomas., Daniela I. Drautz, Arthur M. Lesk, Simon Y. W. Ho, Ji Qi, Aakrosh Ratan, Chih-Hao Hsu, Andrei Sher, Love Dalen, Anders gotherstrom, Lynn P. Tomsho, Snjezana Rendulic, Michael Packard, Paula F. Campos, Tatyana V. Kuznetsova, Fyodor Shidlovsky, Alexei Tikhonov, Eske Willerslev, Paola Iacumin, Bernard Buigues, Per G. P. Ericson, Mietje Germonpre, Pavel Kosintsev, Vladimir Nikolaev, Malgosia Nowak-Kemp, James R. Knight, Gerard P. Irzyk, Clotilde S. Perbost, Karin M. Fredrikson, Timothy T. Harkins, Sharon Sheridan, Webb Miller, and Stephan C. Schuster (2008). "Intraspecific phylogenetic analysis of Siberian woolly mammoths using complete mitochondrial genomes." *Proceedings of the National Academy of Sciences USA* 105(24): 6.

3   Johnson, C. N. (2006). *Australia's Mammal Extinctions: A 50,000 Year History.* Cambridge: Cambridge University Press.

4   Miller, W., Nicola E. Wittekindt, Desiree C. Petersen, et al. (unpublished manuscript). "DiversiTyping the Tasmanian devil: a strategy for selecting insurance populations of endangered species."

5   Stone, R. (2001). *Mammoth: The Resurrection of an Ice Age Giant.* Cambridge, MA: Perseus.

6   Ibid.

7   Nicholls, H. (2008). "Let's make a mammoth." *Nature* 456: 310–14.

8   Campbell, K. L., Jason E. E. Roberts, Laura N. Watson, Jörg Stetefeld, Angela M. Sloan, Anthony V. Signore, Jesse W. Howatt, Jeremy R.H. Tame, Nadin Rohland, Tong-Jian Shen, Jeremy J. Austin, Michael Hofreiter, Chien Ho, Roy E. Weber, and Alan Cooper (2010). "Substitutions in woolly mammoth hemoglobin confer biochemical properties adaptive for cold tolerance." *Nature Genetics* 42: 536–40.

THE STRANGE TALE OF THE NAZI COW

1   Van Vuure, C. (2005). *Retracing the Aurochs: History, Morphology, and Ecology of an Extinct Wild Ox.* Sofia-Moscow: Pensoft.

2   Bunzel-Druke, M. (2001). *Ecological substitutes for Wild Horse (Equus ferus Boddaert, 1785 = E. przewalslii Poljakov, 1881) and Aurochs (Bos primigenius Bojanus, 1827). Natur-und Kulturlandschaft,* Höxter/Jena, 4.

3   Lynch, Anthony H., Julie Hamilton, and Robert E. M. Hedges (2008). "Where the wild things are: aurochs and cattle in England." *Antiquity* 82: 1025–39.

4   Ackerman, D. (2007). *The Zookeeper's Wife.* New York: Norton.

5   Ibid.

6   Heck, L. (1954). *Animals: My Adventure.* London: Methuen.

7   Mysterud, A., K. A. Barton, B. Jedrejewska, Z. A. Krasinski, M. Niedziatkowska, J. F. Kamler, N. G. Yoccoz, and N. C. Stenseth (2007). "Population ecology and conservation of endangered megafauna: the case of the European bison in Bialowieza primeval forest, Poland." *Animal Conservation* 10: 77–87.

8   Ackerman, *The Zookeeper's Wife.*
9   Heck, *Animals.*
10 de Bruxelles, S. (2009). "A shaggy cow story: how a Nazi experiment brought extinct aurochs to Devon." *Times Online*, April. http://www.timesonline.co.uk/tol/news/science/article6143767.ece
11 Clutton-Brock, J. (1999). *A Natural History of Domesticated Mammals.* Cambridge: Cambridge University Press.
12 Guthrie, D. (2005). *The Nature of Paleolithic Art.* Chicago: University of Chicago Press.
13 Bruford, Michael W., Daniel G. Bradley, and Gordon Luikart (2003). "DNA markers reveal the complexity of livestock domestication." *Nature Reviews Genetics* 4: 900–10.
14 Wooding, S. P. (2007). "Following the herd." *Nature Genetics* 39(1): 7–8.
15 Hodder, K., J. M. Bullock, P. C. Buckland, and K. J. Kirby (2005). *Large Herbivores in the Wildwood and Modern Naturalistic Grazing Systems.* English Nature Research Reports 648. Peterborough, UK: English Nature.
16 Klaffke, O. (1999). "The company of wolves." *New Scientist*, February 5.
17 Hodder, Kathy, and James Bullock (2005). "Nature without nurture?" *Planet Earth*, Winter: 30–31.
18 de Bruxelles, "A shaggy cow story."
19 Gillis, A. M. (1991). "Should cows chew cheatgrass on commonlands? Environmentalists and ranchers have beefs about grazing issues." *BioScience* 41(10): 668–75.
20 Knapp, Alan K., John M. Blair, John M. Briggs, Scott L. Collins, David C. Hartnett, Loretta C. Johnson, and E. Gene Towne (1999). "The keystone role of bison in North American tallgrass prairie." *BioScience* 49(1): 39–50.
21 Fuhlendorf, Samuel D., and David M. Engle (2001). "Restoring heterogeneity on rangelands: ecosystem management based on evolutionary grazing patterns." *BioScience* 51(8): 625–32.
22 Kilpatrick, Marm, Colin M. Gillin, and Peter Daszak (2009). "Wildlife-livestock conflict: the risk of pathogen transmission from bison to cattle outside Yellowstone National Park." *Journal of Applied Ecology* 46: 476–85.
23 Towne, E.Gene, David C. Hartnett, and Robert C. Cochran (2005). "Vegetation trends in tallgrass prairie from bison and cattle grazing." *Ecological Applications* 15(5): 1550–59.
24 White, C. (2008). *Revolution on the Range: The Rise of a New Ranch in the American West.* Washington, DC: Island Press/Shearwater Books.
25 Ibid.
26 Stolzenburg, W. (2000). "Good cow, bad cow." *Nature Conservancy*, July/August: 14–19.
27 Curtin, Charles., Nathan F. Sayre, and Benjamin D. Lane (2002). "Transformations of the Chihuahuan borderlands: grazing, fragmentation, and biodiversity conservation in desert grasslands." *Environmental Science and Policy* 5: 55–68.
28 Sayre, N. (2006). *Working Wilderness: The Malpai Borderlands Group and the Future of the Western Range.* Tucson, AZ: Rio Nuevo.
29 Van Auken, O. W. (2000) "Shrub invasions of North American semiarid grasslands." *Annual Review of Ecology and Systematics* 31: 197–215; Allen, C.D. (1998).

"Where have all the grasslands gone?" *Quivera Coalition Newsletter*, Spring/Summer.

30   Curtin et al., "Transformations of the Chihuahuan borderlands."

31   List, Rurik, Gerardo Ceballos, Charles Curtin, Peter J. P. Grogan, Jesus Pacheco, and Joe Truett (2007). "Historic distribution and challenges to bison recovery in the northern Chihuahuan Desert." *Conservation Biology* 21(6): 1487–94; Curtin et al., "Transformations of the Chihuahuan borderlands."

32   Western, D. (2002). *In the Dust of Kilimanjaro*. Washington, DC: Island Press/Shearwater Books; David Western (2009). African Conservation Centre. Personal communication.

33   Western, David, and David Maitumo (2004). "Woodland loss and restoration in a savanna park: a 20-year experiment." *African Journal of Ecology* 42(2): 111–21.

34   Western, David, Rosemary Groom, and Jeffrey Worden (2009). "The impact of subdivision and sedentarization of pastoral lands on wildlife in an African savannah." *Biological Conservation* 142(11): 2538–46.

35   Mioron, Joseph S., Joseph Lemomo Munge, Dennis Ole Sonkoi, Periperi Toroke, Jackson Lowasa Ole Teteyo, and Yusuf Ole Petenya (2004). "Press statement on a visit of Kenyan and Tanzanian Maasai to the Malpai Borderlands Group, Arizona-New Mexico." Available from African Conservation Centre, http://www.conservationafrica.org/en/

36   Curtin, Charles, and David Western (2008). "Grasslands, people, and conservation: over-the-horizon learning exchanges between African and American pastoralists." *Conservation Biology* 22(4): 870–77.

37   Western, Personal communication.

38   Wolf, Benjamin, Xunhua Zheng, Nicolas Brüggemann, Weiwei Chen, Michael Dannenmann, Xingguo Han, Mark A. Sutton, Honghui Wu, Zhisheng Yao, and Klaus Butterbach-Bahl (2010). "Grazing-induced reduction of natural nitrous oxide release from continental steppe." *Nature* 464: 881–84.

EPILOGUE: THE GIANTS THIS TIME

1   Barnosky, A. D. (2008). "Megafauna biomass tradeoff as a driver of Quaternary and future extinctions." *Proceedings of the National Academy of Sciences USA* 105(suppl. 1): 11543–48.

2   McDaniel, Carl N., and David N. Borton (2002). "Increased human energy use causes biological diversity loss and undermines prospects for sustainability." *BioScience* 52(10): 929–36.

INDEX